EQUILIBRIUM THERMODYNAMICS IN PETROLOGY

AN INTRODUCTION

ROGER POWELL

EQUILIBRIUM THERMODYNAMICS IN PETROLOGY

AN INTRODUCTION

ROGER POWELL

Department of Earth Sciences
University of Leeds
Leeds, England

Harper & Row, Publishers

London   New York   Hagerstown
San Francisco .   Sydney

British Library Cataloguing in Publication Data

Powell, Roger
    Equilibrium thermodynamics in petrology.
    1. Petrology   2. Thermodynamics
    I. Title
    552'.001'5367   QE431.6.T45

        ISBN 0–06–318061.8
        ISBN 0–06–318073.1 Pbk

Designed by Richard Dewing 'Millions'
Typeset by Universities Press (Belfast) Ltd
Printed by Butler & Tanner Ltd, Frome and London

Dedicated to Marge

Contents

Preface

The basic premise in this book is that at least some of the features in many rocks can be interpreted as the result of the achievement of equilibrium on some scale at some time or times during their evolution. Given this premise, equilibrium thermodynamics provides a way of looking at rocks, not only for discovering at what conditions they formed, for example the temperature and pressure of formation, but also for understanding the processes involved in their formation.

The purpose of this book is to provide an introduction to the methods of equilibrium thermodynamics via an axiomatic approach which concentrates on those equations which are most useful in petrology, which also results in a development which is mathematically straightforward. Equilibrium thermodynamics, in its quantitative form of equations and numbers, as well as its qualitative form of phase diagrams, is a very practical subject, so a crucial part of this book is the abundant worked examples which cover a range of the petrological applications of equilibrium thermodynamics. There is an emphasis on metamorphic and igneous applications because there is more evidence for the attainment of mineralogical equilibrium in these types of rocks.

This book is intended primarily as a final year undergraduate and first year postgraduate text to be used in geochemistry and petrology courses. The book should also be useful for research workers in petrology who are unfamiliar with the application of equilibrium thermodynamics in petrology.

Chapters 2 and 3 provide an introduction to equilibrium
thermodynamics, along with an introduction to the different types of
phase diagrams which are important in summarizing equilibrium phase
relations and in the qualitative consideration of equilibrium in rocks.
Aspects of the information required for thermodynamic calculations are
considered in chapters 4–6. In chapters 7–8, the emphasis is on the
processes by which the mineralogy and compositions of rocks change,
mainly using phase diagrams. Chapter 9 comprises a series of
petrological topics illustrating ideas about geological processes which
can be obtained by qualitative equilibrium considerations of rocks or
groups of rocks. The references at the end of each chapter include
essential sources and some suggestions for further and advanced
reading, which are intended to allow entry into the recent literature.

I would like to thank Drs A. Bath, R. A. Cliff, C. Graham, D. Hirst,
R. K. O'Nions, M. Powell and S. W. Richardson for critically
reviewing an earlier draft of this book. Conversations over the last
several years with Professor I. S. E. Carmichael, Drs R. K. O'Nions,
S. W. Richardson, M. Norry, M. Bickle, and, of course, M. Powell
have contributed to the emergence of this book. The patience and help
of my wife, Marge, cannot be overestimated. Of course, any factual
and typographical errors are mine.

Chapter 1

Introduction

The Earth and each of its constituent parts can be treated as chemical systems. The term, system, is used to define the amount of matter being considered—it may be a small volume of a mineral, a rock or even the whole Earth, depending on the scale of the processes being considered. The system will consist of one or more phases (minerals, gases and liquids). The boundaries of our system are not considered to be barriers to energy (so that, for example, temperature differences across the boundary can be eliminated), or to the movement of materials (elements, a fluid phase, and so on) across the boundary (so that composition differences across the boundary can be eliminated). It may turn out that the boundaries of our system are barriers to these processes, particularly the latter, depending on the time during which the processes can operate. A system is closed if the boundary of the system has acted as a barrier to the transfer of material, and is open if the boundary has not acted as a barrier. During the metamorphism of a basalt to a greenschist, a hand-specimen-sized system may have remained a closed system to many major elements (Si, Al, Ca etc.) but was open to some trace elements (e.g. Rb) and the fluid phase (H_2O and CO_2). The behaviour of our system of interest may change dramatically during its development through time.

Equilibrium is a concept which is easily defined for simple chemical systems. For example, if we hold a mechanical mixture of halite (NaCl) and sylvite (KCl) at 300 °C and 1 bar indefinitely, and measure the compositions of the phases at regular intervals, the two phases would gradually dissolve in each other until the phases stopped changing composition with time. The phases have reached equilibrium—their compositions would not change if they were left indefinitely at these conditions. Note that if the conditions are changed then the phases will gradually change composition until they are in equilibrium at the new conditions. A crucial aspect of equilibrium is the rate at which equilibrium is attained, in this case, it will depend on the rate of migration and incorporation of Na from NaCl into KCl and K from KCl into NaCl. Clearly these rates are going to be dependent on many things, for example grain size and the presence of an interstitial fluid phase—as well as temperature and pressure.

Geological materials are not the result of well controlled chemical experiments. For example, during the metamorphism of a pelite, the original clay or silt follows some temperature–pressure (P–T) path through time (t) as a result of burial and heating followed by uplift and cooling. Can we talk about equilibrium in the metamorphic process? This will depend almost completely on the rate at which equilibrium is attained along the P–T–t path. If the rate of equilibration is fast compared to the rate at which P and T are changing with time, then the rock will maintain equilibrium throughout its P–T path and would reach the Earth's surface as some low temperature assemblage of clay minerals, etc. This does not usually happen, the rock reaches the Earth's surface with a high P–T assemblage. A reasonable interpretation is that the rate of attainment of equilibrium became much slower than the rate of change of P–T with time during cooling, so that the mineralogy of the rock records some high P–T equilibrium attained before substantial cooling and uplift took place. The equilibrium represented in the mineralogy has been *frozen in.*

Metamorphic rocks are often more complicated, for example, the minerals may be zoned, the rock may have been metamorphosed several times, or late-stage alteration minerals may be developed. Zoning can be interpreted in several ways, but primarily it indicates that the rate of formation of unzoned grains was slower than the rate of change of P–T with time. For example, a zoned garnet in a metapelite may have formed as the result of growth of the garnet over a section of the P–T–t path, with each new zone of garnet attaining equilibrium with the rest of the minerals while the rate of change of composition of previously-formed zones in the garnet was insufficient to form homogeneous grains. The presence of late-stage alteration minerals suggests that the high P–T assemblage attempted to equilibrate at new lower P–T conditions under certain favourable circumstances. In the case of the late-stage replacement of garnet by chlorite, the favourable circumstance may be the influx of water into the assemblage. In the case of mineral zoning, each new zone on the surface of the mineral may be in equilibrium with the rest of the assemblage, so that we may be able to refer to surface equilibrium between the mineral and the rest of the assemblage. In the case of alteration minerals, equilibrium may be attained *between* the alteration minerals, but not between the alteration minerals and the minerals they alter.

The same arguments can be developed for igneous and sedimentary rocks; certain features of the assemblages can be interpreted in terms of equilibrium attained at different times under different conditions. It should be clear that equilibrium may be achieved or nearly achieved at various stages in the evolution of a rock, with remnants of these equilibria frozen in the rock. Therefore, for each rock, the crux is the *geological* interpretation of the various equilibria which might have been achieved during the evolution of the rock; whether some minerals were only in surface equilibrium with the rest of the assemblage, which minerals formed later than the main assemblage, and so on.

Two aspects of the achievement of equilibrium at a particular stage in the evolution of a rock are:

1   the time available for equilibrium to be achieved at a particular temperature and pressure,
2   the scale on which equilibrium is achieved.

The longer a system suffers particular conditions, the larger the volume of the system which is able to reach equilibrium. After a short time, a system may consist of many small volumes (domains) which have reached equilibrium, but the system as a whole is out of equilibrium. After a long time, the whole system may have reached equilibrium. This idea of local or mosaic equilibrium in a large system which has not reached equilibrium is important. It allows consideration of equilibrium in many cases where equilibrium on a large scale has not been reached. For example, the biotite and chlorite grains in a thin section of a rock may show a wide range of compositions, but adjacent biotite and chlorite grains may have reached (local) equilibrium. The onus on the petrologist is to recognize the scale on which equilibrium may have been achieved.

Although in the above the achievement of equilibrium in natural processes has been emphasized, it is important to realize that disequilibrium will occur, where a process has taken place which changes the assemblage but the result of the change is still a long way from equilibrium (this is different from the case, for example, where a high $P-T$ assemblage is out of equilibrium at the Earth's surface with no mineralogical or chemical change having occurred there). This may be true particularly at low temperatures, where the rate of equilibration is very slow (for example, many sedimentary processes), and when conditions are changing very quickly, the rate of change of conditions being quicker than the rate of equilibration (for example, in the crystallization of a rapidly cooled lava, in many sedimentary processes, and so on).

Chapter 2

Introduction to Equilibrium Thermodynamics

The methods of equilibrium thermodynamics which are introduced in this chapter can be used to discover the conditions of formation of mineral assemblages which are interpreted to have formed at equilibrium. The types and features of phase diagrams will also be introduced, mainly in the worked examples (WE). The understanding of phase diagrams is essential to the pictorial representation and qualitative consideration of equilibrium in rocks.

Gibbs energy and equilibrium

The classical starting point in equilibrium thermodynamics is a consideration of the so called laws of thermodynamics. However, the derivation of geologically useful equations from these laws is unnecessary here. It can be shown that the equilibrium condition for a system is the minimization of one of several energies; which one depending on the type of system being considered. In much of geology (and chemistry), systems are considered in which temperature and pressure are features of the environment of the system—they can be considered as being constant and superimposed on the system. In these cases, the Gibbs energy is minimized at equilibrium. Systems for which this equilibrium principle holds are examined in this book.

Therefore we will take as our starting point that the Gibbs energy of a system is minimized at equilibrium. Our system may consist of several phases. The Gibbs energy of the system is the sum of the Gibbs energies of the phases in proportion to the amounts of the phases present in the system. Consider figure 2.1, a Gibbs energy–composition $(G-x)$ diagram involving two end-members 1 and 2 and two phases A and B. The Gibbs

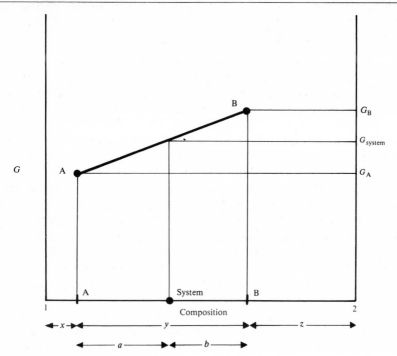

2.1   Gibbs energy–composition diagram between end-members 1 and 2 involving phases A and B showing the additivity of Gibbs energies. The proportion of 1 in A, the mole fraction, $x_{1A} = (y+z)/(x+y+z)$. Similarly $x_{2A} = x/(x+y+z) = 1 - x_{1A}$, $x_{1B} = z/(x+y+z)$, $x_{2B} = (x+y)/(x+y+z) = 1 - x_{1B}$.

energy of A is G_A, of B is G_B. The proportion of A and B in the system is given by the lever rule (see WE2(a)):

$$\text{proportion of A} = \frac{b}{a+b}$$

$$\text{proportion of B} = \frac{a}{a+b}$$

The Gibbs energy of the system is therefore:

$$G_{\text{system}} = G_A \frac{b}{a+b} + G_B \frac{a}{a+b}$$

and is at the appropriate composition along a straight line joining A and B in the diagram. The Gibbs energy of a phase is a measureable property depending on temperature (T), pressure (P), composition and structure of the phase, as well as the amount of phase. It is often convenient to consider the properties (for example Gibbs energy) of a fixed amount of phase. This amount is the mole or gram formula weight involving Avogadro's number ($N = 6 \cdot 0226 \times 10^{23}$) of molecules of the phase. Thus when we refer to the Gibbs energy (G) of a phase, we mean the molar Gibbs energy of the phase.

The simplest application of the principle of the minimization of Gibbs energy is in the consideration of the stability of polymorphs. Consider the $CaCO_3$ polymorphs, calcite, the hexagonal low pressure form, and aragonite, the denser orthorhombic high pressure form. Figure 2.2(a) shows how the Gibbs energies of calcite and aragonite vary with pressure at a particular temperature. On this diagram, calcite is stable with respect to aragonite below 4 kbar because it has a lower Gibbs energy than aragonite under these conditions. On the other hand, aragonite is metastable with respect to calcite under these conditions. Above 4 kbar, aragonite is stable with respect to calcite, while calcite is metastable. Calcite and aragonite can coexist in equilibrium at the pressure of the intersection of the Gibbs energy lines. The essential information from figure 2.2(a), that calcite is stable below 4 kbar and aragonite is stable above 4 kbar at 300 °C can be transcribed onto a pressure–temperature (P–T) diagram, figure 2.2(b). The intersection of the G lines for calcite and aragonite at a range of temperatures maps out a line on the P–T diagram. The slope of the P–T line is controlled by the relative change of the Gibbs energies of calcite and aragonite with temperature and pressure. This line is often referred to in terms of the reaction which takes place across it:

$$CaCO_3 = CaCO_3$$
calcite   aragonite

Note that the equilibrium condition for this reaction, i.e. for the stable coexistence of calcite and aragonite is:

$$G_{CaCO_3, \text{ calcite}} = G_{CaCO_3, \text{ aragonite}}$$

which is the reaction written in terms of Gibbs energies. Given thermodynamic data for calcite and aragonite, this equation allows the calculation of the position of the P–T line along which calcite and aragonite may coexist.

In figure 2.2(b), the stable phase is marked in each field of the P–T diagram. Consider a carbonate-bearing rock metamorphosed at a, it should contain aragonite. This rock is now uplifted to the Earth's surface. The aragonite may not have time to convert to calcite in the stability field of calcite, so we can find aragonite preserved metastably at the Earth's surface although calcite is the stable $CaCO_3$ polymorph under these conditions. Many carbonate-secreting organisms actually grow aragonite shells metastably in the stability field of calcite. These shells get converted to calcite during diagenesis. It is important to realize that a diagram like figure 2.2(b) only contains information on the stable assemblages in each

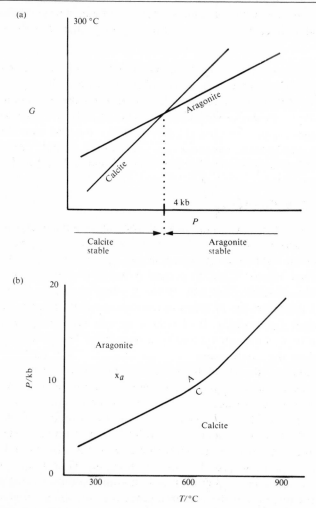

2.2   Schematic Gibbs energy–pressure diagram, (a), and pressure–temperature diagram, (b), for the polymorphic reaction calcite = aragonite in the $CaCO_3$ system.

$P-T$ field. Metastable relations which are most important in understanding geological materials can be visualized only with the help of the Gibbs energy diagrams from which the phase diagram has been constructed.

The consideration of equilibrium becomes more complicated if the phases

are solutions, for example, if we want to know the $P-T$ line for equilibrium between calcite and aragonite of particular compositions in the
system $CaCO_3-SrCO_3$. Here we need to examine Gibbs energy–
composition $(G-x)$ diagrams and relate them to temperature–composition
$(T-x)$ and $P-T$ diagrams. The form of the phase diagram is now controlled not only by the way the Gibbs energies of the phases change with
pressure and temperature, but also the way they change with composition.

The compositional dependence of the Gibbs energy of a phase is strongly
influenced by the amount of possible solution between the end-members
of the phase. If there is complete solution between the end-members, as
in for example $Fe_2SiO_4-Mg_2SiO_4$ olivines, the $G-x$ curve is open,
whereas if there is only limited solution between the end-members, as in
for example $Al_2O_3-SiO_2$ where there is very limited solubility of Al_2O_3 in
quartz, the $G-x$ curve is tightly compressed (figures 2.3(a) and (b)).

$G-x$ and $T-x$ diagrams

Gibbs energy–composition diagrams for systems at a particular temperature and pressure can be used to show some of the aspects of the
minimization of Gibbs energy principle for equilibrium. Consider figure
2.4, $G-x$ diagrams for a binary system between end-members 1 and 2,
involving three possible phases A, B and C—two of which (A and C) can
show only a limited range of solution towards the other phases, hence the
sharply curved $G-x$ curves. B can show an extended range of solution
towards the two end-members and so has an open $G-x$ curve. A simple
way of finding the equilibrium phase assemblage for a particular composition system is shown in figure 2.4. Suppose we wish to find the
equilibrium assemblage of phases and their compositions for a system of
composition, f. Place a ruler on the x-axis on the diagram and allow the
ruler to pivot at f. Now, taking the $G-x$ loops for each phase as solid
lobes which the ruler cannot cross, the ruler is moved up the diagram, still
pivoted at the composition of f, until the ruler cannot be moved further.
The points or point where the ruler is touching $G-x$ loops gives the
equilibrium assemblage of phases and their compositions. In this case, the
ruler fixes the position of the common tangent to the $G-x$ loops of A and
C. In figure 2.5(a), all three compositions f, g and h give A+C as the
equilibrium assemblage, the compositions of A and C being the same in
each case, but the proportions are different, f containing about 75% of A
and 25% of C, and so on. In (b), the equilibrium assemblage of f is A+B,
of g is B, and of h is B+C; the compositions of the phases being given by
the position of the tangents to the $G-x$ curves. Note that the composition
of B is quite different for the three compositions, f, g and h.

A useful way of representing the equilibrium information on a $G-x$

diagram, for every composition in a system, is by using a compatibility diagram (figures 2.5(c) and (d)). Bars are used to represent the extent of solution of the phases. The compositions of the phases in the two-phase fields are given by the compositions of the ends of the bars adjacent to the two-phase fields. Compatibility diagrams for a range of temperatures can be combined to give a more familiar temperature–composition (T–x) diagram, for example figure 2.6. At the temperature of E in the diagram,

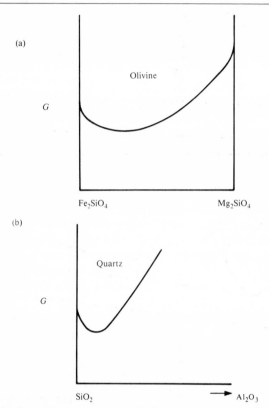

2.3   Schematic G–x diagrams showing in (a) an open loop reflecting the 'easy' substitution of Fe for Mg in olivine, in (b) a tight loop reflecting the 'difficult' substitution of Al_2O_3 in quartz.

the tangent for intermediate compositions touches the G–x loops of A, B and C. This diagram is similar to the melting diagrams in many binary systems; for example A = diopside, B = liquid and C = anorthite, E on the diagram is the eutectic.

It is constructive to consider metastability in this type of system, with the help of figure 2.5. Imagine that, under the conditions of figure 2.5(b) for composition f, phase B will not nucleate and grow (say during the duration of an experiment). Nevertheless, the system can still reach equilibrium, although this is now a metastable equilibrium. The system

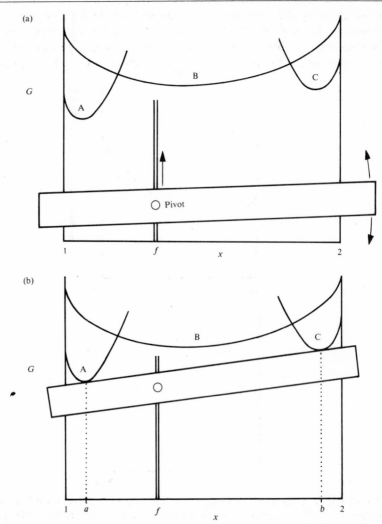

2.4 A method for finding the equilibrium assemblage and phase compositions for a particular composition on a $G–x$ diagram using a ruler pivoted at the composition of interest, f, and allowed to move up the diagram with the pivot placed in a slot. Raising the ruler from its position in (a) until it abuts the $G–x$ loops for A and C in (b) gives the equilibrium assemblage for composition, f, of phase A of composition a and phase C of composition b.

will contain A + C metastably with respect to A + B, the compositions of A and C at equilibrium being given by the position of the common tangent to A and C, ignoring the $G–x$ loop of B. It is helpful to include the metastable extensions of boundaries on $T-x$ diagrams to show the compositions of the phases in metastable equilibrium (broken lines in figure 2.6).

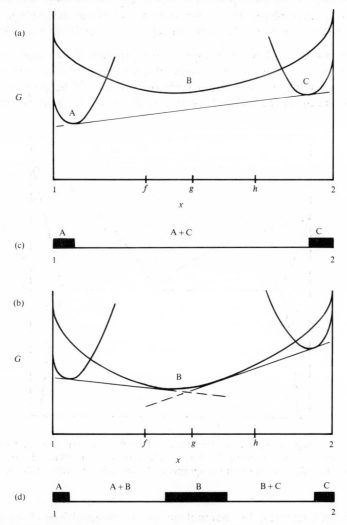

2.5 $G-x$ diagrams at two temperatures, (a) and (b), showing the effect of the change in the relative position of the $G-x$ loop for B with respect to the $G-x$ loops for A and C on the equilibrium assemblages, reflected in the compatibility diagrams, (c) and (d).

This points to a fundamental feature of equilibrium thermodynamics—it will provide us with the equilibrium assemblage and the equilibrium compositions of the phases for a particular composition system for the phases which we say can occur in the system. If we leave out a phase, then thermodynamics cannot tell us that we should have included it. In figure 2.5(b), for f, A+B is the stable equilibrium assemblage, while A+C is the metastable equilibrium assemblage. If we then discover that there is another phase, D, in our system, our stable assemblage A+C might in fact be metastable with respect to an assemblage involving D.

Bearing this in mind, we can appreciate a problem in the application of equilibrium thermodynamics to mineral assemblages. Given a particular mineral assemblage, equilibrium thermodynamics will be able to tell us the conditions at which that system could have been in equilibrium

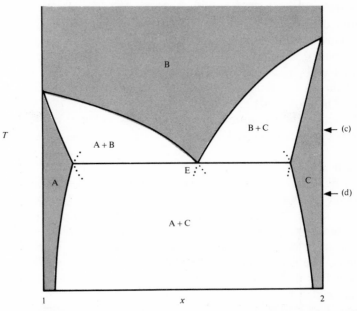

2.6 A $T-x$ diagram which can be generated from $G-x$ diagrams for a series of temperatures including those in figure 2.5. The sections marked (c) and (d) correspond to the compatibility diagrams in figure 2.5. The one-phase fields are shaded.

(naively, by comparing the assemblage and the compositions of the phases with compatibility diagrams for known pressures and temperatures), not whether equilibrium is actually reasonable for that assemblage. The geological interpretation of equilibrium is always the fundamental step in the application of equilibrium thermodynamics to rocks.

Figures 2.5 and 2.6 show how a well known type of phase diagram is related to the relative positions and shapes of $G-x$ loops for the phases. Clearly, different phase diagrams are generated with other relative positions and shapes.

The stable assemblages in more complicated systems can be obtained by minimizing the Gibbs energy of the system in the same graphical way as was considered in figure 2.4, although it may be difficult to visualize let alone portray the $G-x$ diagrams. For these more complicated systems and even for the simple systems it would be more convenient to be able to determine the phase diagrams (or the compositions of coexisting phases at equilibrium) from algebraic expressions for the $G-x$ loops of the phases.

Algebraic equilibrium relations

There are two ways of arriving at the algebraic equilibrium relations for a system—both of which require some mathematical manipulation and calculus (refer to appendix C for any necessary revision!). Both methods are outlined below for equilibrium between two phases A and B in a binary system involving end-members 1 and 2. We can represent the compositions of the phases in two ways; firstly in terms of the number of moles of 1 and 2 in A, n_{1A} and n_{2A}, and in B, n_{1B} and n_{2B}; secondly in terms of the mole fractions of 1 and 2 in A, x_{1A} and x_{2A}, and in B, x_{1B} and x_{2B}. The mole fractions are proportions, thus $x_{1A} = n_{1A}/(n_{1A} + n_{2A})$. Note that $x_{1A} + x_{2A} = 1$, so that for example $x_{2A} = 1 - x_{1A}$. G_A and G_B are the Gibbs energies of phases A and B.

Approach 1
Consider figure 2.7. The equilibrium compositions of coexisting A and B involves the position of the common tangent to the $G-x$ loops of A and B—ZY in figure 2.7. We want to formulate this equilibrium statement mathematically. Consider any line which is tangent to the $G-x$ loop for A, for example mn; and a line which is tangent to the other loop, for example rs. The line which fixes the equilibrium compositions for coexisting A and B is the common tangent to the two loops which is produced when mn and rs are made to be coincident by being moved round their respective $G-x$ loops. In this case, the intercept on the G-axis where $x_1 = 0$ is at Z for both mn and rs, and the intercept on the G-axis where $x_1 = 1$ is Y for both mn and rs. The tangent to a curve can be formulated using differentiation. Using the definition of differentiation involving gradients (p. 260):

$$\frac{\mathrm{d}G_A}{\mathrm{d}x_1} = \frac{G_Y - G_A}{1 - x_{1A}} \quad \text{and} \quad \frac{\mathrm{d}G_B}{\mathrm{d}x_1} = \frac{G_Y - G_B}{1 - x_{1B}} \tag{2.1}$$

$$\frac{\mathrm{d}G_A}{\mathrm{d}x_2} = \frac{G_Z - G_A}{1 - x_{2A}} \quad \text{and} \quad \frac{\mathrm{d}G_B}{\mathrm{d}x_2} = \frac{G_Z - G_B}{1 - x_{2B}} \tag{2.2}$$

Re-arranging (2.1):

$$G_Y = G_A + (1 - x_{1A})\frac{\mathrm{d}G_A}{\mathrm{d}x_1} \quad \text{and} \quad G_Y = G_B + (1 - x_{1B})\frac{\mathrm{d}G_B}{\mathrm{d}x_1} \tag{2.3}$$

Re-arranging (2.2):

$$G_Z = G_A + (1 - x_{2A}) \frac{dG_A}{dx_2} \quad \text{and} \quad G_Z = G_B + (1 - x_{2B}) \frac{dG_B}{dx_2} \qquad (2.4)$$

Combining (2.3) and (2.4):

$$G_A + (1 - x_{1A}) \frac{dG_A}{dx_1} = G_B + (1 - x_{1B}) \frac{dG_B}{dx_1}$$

$$G_A + (1 - x_{2A}) \frac{dG_A}{dx_2} = G_B + (1 - x_{2B}) \frac{dG_B}{dx_2} \qquad (2.5)$$

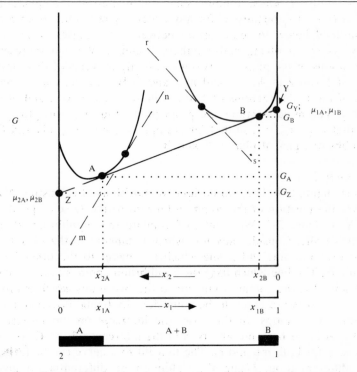

2.7 A G–x diagram illustrating the derivation of the algebraic statement of the principle of the minimization of Gibbs energy at equilibrium given in the text.

These two equations are the two equilibrium relations for the coexistence of A and B giving the compositions of the two phases. These compositions can be calculated by solving the two equations for the two unknowns, x_{1A} ($= 1 - x_{2A}$) and x_{1B} ($= 1 - x_{2B}$), if G_A and G_B are expressed as functions of x_1 and x_2 so that the differentials can be evaluated.

Now, another differential of the Gibbs energy with respect to composition is often used in thermodynamics, the chemical potential, μ. In this example, the chemical potential of component 1 in phase A, μ_{1A}, is defined as:

$$\mu_{1A} = \left(\frac{\partial(n_{1A} + n_{2A})G_A}{\partial n_{1A}}\right)_{n_{2A}}, \quad \text{also} \quad \mu_{2A} = \left(\frac{\partial(n_{1A} + n_{2A})G_A}{\partial n_{2A}}\right)_{n_{1A}} \quad (2.6)$$

and

$$\mu_{1B} = \left(\frac{\partial(n_{1B} + n_{2B})G_B}{\partial n_{1B}}\right)_{n_{2B}}, \quad \text{also} \quad \mu_{2B} = \left(\frac{\partial(n_{1B} + n_{2B})G_B}{\partial n_{2B}}\right)_{n_{1B}} \quad (2.7)$$

or, in words, the rate of change of the Gibbs energy of an arbitrary amount of phase (hence the Gibbs energy of the phase is multiplied by the total amount of the phase, $n_{1A} + n_{2A}$ for A and $n_{1B} + n_{2B}$ for B) with respect to one of the amounts of an end-member, keeping the other amounts constant. Each of these chemical potentials can be rewritten in terms of the gradients on figure 2.7. Thus by differentiating $(n_{1A} + n_{2A})G_A$ with respect to n_{1A} by parts:

$$\mu_{1A} = \left(\frac{\partial(n_{1A} + n_{2A})G_A}{\partial n_{1A}}\right)_{n_{2A}} = G_A + \left(\frac{\partial G_A}{\partial n_{1A}}\right)_{n_{2A}} \quad (2.8)$$

Differentiating with respect to x_{1A} rather than n_{1A}:

$$\mu_{1A} = G_A + \frac{dG_A}{dx_{1A}} \cdot \left(\frac{\partial x_{1A}}{\partial n_{1A}}\right)_{n_{2A}} = G_A + (1 - x_{1A})\frac{dG_A}{dx_{1A}} \quad (2.9)$$

Comparing this result with (2.3) shows that $\mu_{1A} = G_Y$. By similar reasoning $\mu_{1B} = G_Y$, and $\mu_{2A} = \mu_{2B} = G_Z$. Thus the equilibrium conditions for coexisting A and B (2.5) can be rewritten as:

$$\mu_{1A} = \mu_{1B}$$
$$\mu_{2A} = \mu_{2B} \quad (2.10)$$

The chemical potentials of the various end-members of a phase can be visualized in terms of the tangent to the $G-x$ surface at that composition. The chemical potential of each end-member for that composition is given by the Gibbs energy on the tangent where the mole fraction of that end-member is equal to one, i.e. it is the sole constituent of the phase. Therefore, the chemical potential of an end-member in a phase consisting of just that end-member is equal to the Gibbs energy of the phase.

Approach 2
The above approach provides a clear connection between the graphical and algebraic approaches to equilibrium for binary systems, but would be difficult to develop for more complicated systems. The second approach is more difficult but is easily generalizable to larger systems—it is based on

the calculus representation of the condition that the Gibbs energy of the system is minimized at equilibrium. Consider coexisting phases A and B at equilibrium as in the above example. If we alter the composition of A from its equilibrium value this will increase the Gibbs energy of the system, because the Gibbs energy of the system had its minimum value with A (and B) having the equilibrium composition. Thus, if we plot the Gibbs energy of the system against any composition term, for example n_{1A}, keeping all other composition terms constant, we would get a diagram like figure 2.8.

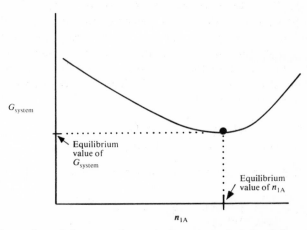

2.8 An illustration of the second approach to the principle of the minimization of Gibbs energy at equilibrium, involving $dG_{system}/dn_{1A} = 0$ at equilibrium.

Thus the equilibrium value of n_{1A} is given by:

$$\left(\frac{\partial G_S}{\partial n_{1A}}\right)_{n_{2A}, n_{1B}, n_{2B}} = 0 \tag{2.11}$$

or, in words, when this tangent to the Gibbs energy of the system curve is horizontal, i.e. when the gradient is zero. Now, the Gibbs energy of the system is:

$$G_S = (n_{1A} + n_{2A})G_A + (n_{1B} + n_{2B})G_B \tag{2.12}$$

Applying (2.11) to (2.12) we have at equilibrium:

$$0 = \left(\frac{\partial(n_{1A} + n_{2A})G_A}{\partial n_{1A}}\right)_{n_{2A}, n_{1B}, n_{2B}} + \left(\frac{\partial(n_{1B} + n_{2B})G_B}{\partial n_{1A}}\right)_{n_{2A}, n_{1B}, n_{2B}} \tag{2.13}$$

We can change the variable we are differentiating with respect to using $n_1 = n_{1A} + n_{1B}$, the total amount of end-member 1 in the system. Thus:

$$\left(\frac{\partial}{\partial n_{1A}}\right)_{n_{2A}, n_{1B}, n_{2B}} = -\left(\frac{\partial}{\partial n_{1B}}\right)_{n_{1A}, n_{2A}, n_{2B}}$$

Applying this sleight of hand to the second term in (2.13):

$$\left(\frac{\partial(n_{1A}+n_{2A})G_A}{\partial n_{1A}}\right)_{n_{2A},n_{1B},n_{2B}} = \left(\frac{\partial(n_{1B}+n_{2B})G_B}{\partial n_{1B}}\right)_{n_{2B},n_{1A},n_{2A}}$$

Comparing with (2.6a) and (2.7a), this is:

$$\mu_{1A} = \mu_{1B}$$

By employing the same logic on the equivalent of (2.11) where the differentiation is with respect to n_{2A}, we get:

$$\mu_{2A} = \mu_{2B}$$

and these two equations, which are the same as (2.10), are the equilibrium relations for the system.

General equilibrium relation

In the above two sections we have discovered that the conditions for the equilibrium coexistence of phases A and B in this binary system are:

$$\mu_{1A} = \mu_{1B} \quad \text{and} \quad \mu_{2A} = \mu_{2B}$$

They correspond to the two balanced chemical reactions:

end-member 1 in A = end-member 1 in B

and

end-member 2 in A = end-member 2 in B

written in terms of chemical potentials. This correspondence between chemical reactions and chemical potential equations is quite general. The equilibrium relation for any balanced chemical reaction, usually written between end-members of the phases, is the same reaction written in terms of chemical potentials. This can be written as

$$\Delta\mu = 0$$

where Δ is used to represent the change of a property for a balanced reaction, meaning (sum of properties of products) − (sum of properties of reactants).

For example, one of the conditions for equilibrium between clinopyroxene (cpx), plagioclase (pl) and quartz (q) in, for example, a blueschist is:

$$\mu_{SiO_2,q} + \mu_{NaAlSi_2O_6,cpx} = \mu_{NaAlSi_3O_8pl}$$

which corresponds to the reaction:

$$SiO_2 + NaAlSi_2O_6 = NaAlSi_3O_8$$
$$\quad q \qquad\quad cpx \qquad\qquad pl$$

If this were written in the form $\Delta\mu = 0$, then:

$$\Delta\mu = \mu_{NaAlSi_3O_8,pl} - \mu_{SiO_2,q} - \mu_{NaAlSi_2O_6,cpx} = 0$$

The end-members in the reaction need not all be in different phases. For example, the equilibrium composition of a carbon–hydrogen–oxygen (C–H–O) gas, in terms of the amounts of the different species (i.e. end-members) in the gas are determined by the relations:

$$\mu_{CO_2} = \mu_{CO} + \tfrac{1}{2}\mu_{O_2} \qquad \text{corresponding to} \quad CO_2 = CO + \tfrac{1}{2}O_2$$

$$\mu_{O_2} + \mu_{CH_4} = \mu_{CO_2} + 2\mu_{H_2} \quad \text{corresponding to} \quad O_2 + CH_4 = CO_2 + 2H_2$$

and so on, where each chemical potential refers to that end-member of the gas.

If we are thinking about the frozen-in equilibria in a rock then there are going to be many equilibrium relations we can write, each one corresponding to a different chemical reaction between the end-members of the minerals, not only for reactions which involve end-members that make up a substantial part of a phase, but also for reactions which involve end-members that are only present in a phase in trace amounts. For example, in the blueschist example above, it would be reasonable to consider the equilibrium relation:

$$\mu_{SiO_2,q} + \mu_{SrAl_2SiO_6,cpx} = \mu_{SrAl_2Si_3O_8,pl}$$

even though the amounts of Sr in the clinopyroxene and plagioclase are very small indeed.

We can also write equilibrium relations involving phases no longer present in our rock, for example the metamorphic fluid phase or the magma which deposited a cumulate. For example, the equilibrium relation for:

$$2MgSiO_3 = Mg_2SiO_4 + SiO_2$$

$$\quad\text{pyroxene} \qquad \text{olivine} \qquad \text{magma}$$

is

$$2\mu_{MgSiO_3,px} = \mu_{Mg_2SiO_4,ol} + \mu_{SiO_2,magma}$$

Equilibrium relations involving end-members of phases now absent allow us to consider what these phases were like, for example the composition of a metamorphic fluid phase.

The real constraint on the consideration of equilibrium for reactions between end-members of minerals is whether we can actually calculate the chemical potentials involved. The formulation of chemical potentials takes up much of the next chapter, as much of the application of equilibrium thermodynamics in geology depends on this.

Worked examples 2

2(a) Plot the positions of forsterite (FO) and enstatite (EN) on a line representing the binary system MgO–SiO_2.

Here we use the lever rule in reverse because we know the mole proportion of MgO in forsterite is $\frac{2}{3}$ and the proportion of SiO_2 is $\frac{1}{3}$ from the formula of forsterite, Mg_2SiO_4. Therefore, if we divide the x-axis into three equal parts, forsterite plots on the mark nearer MgO, figure 2.9(a).

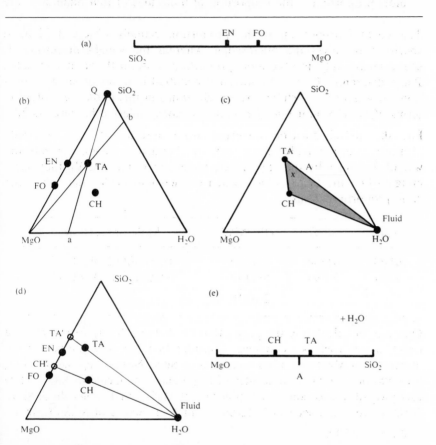

2.9 The positions of enstatite (EN) and forsterite (FO) on a MgO–SiO_2 diagram are shown in (a). (b) shows the positions of talc (TA) and chrysotile (CH) on a MgO–SiO_2–H_2O diagram, with the construction lines for plotting the position of TA. (c) illustrates the use of a triangular diagram as a compatibility diagram. (d) shows the position of TA and CH on MgO–SiO_2 after projection from H_2O. (e) is the MgO–SiO_2 compatibility diagram which corresponds to (c) after projection from H_2O. (WE 2(a) and 2(c))

For enstatite, the proportions of MgO and SiO_2 are both $\frac{1}{2}$, so enstatite plots centrally on the axis.

2(b) An analysis of a rock consisting of olivine (Mg_2SiO_4) and orthopyroxene ($MgSiO_3$) gives MgO 45·06 wt% and SiO_2 54·96 wt%. What are the proportions of the minerals in the rock?

Proportion of a mineral can be interpreted in three ways:
  weight proportion is the weight of that mineral in a unit weight of rock,
  volume proportion is the volume of that mineral in a unit volume of rock,
  mole proportion is the proportion of molecules of that mineral in the rock.

The volume proportion is the proportion actually observed in hand specimen or down the microscope, whereas the weight proportion is useful analytically and the mole proportion is useful thermodynamically. Phase diagrams $(T - x$, etc.) are usually plotted in terms of mole proportions or weight proportions. All the diagrams in this book are plotted in terms of mole proportions (i.e. mole fractions) unless otherwise stated.

First the analysis must be converted into a more useful form. The mole proportions of the oxides in the rock are obtained by dividing each of the weight % values by the appropriate molecular weight (MW), then summing to find the factor which, when divided into each value, gives the mole proportion. Thus:

Oxide	wt %	wt%/MW	Mole proportion
MgO	45·06	1·1176	$1·1176/2·0319 = 0·55$
SiO_2	54·94	0·9143	$0·9143/2·0319 = 0·45$
		2·0319	

One way of calculating the proportion of olivine (ol) and orthopyroxene (opx) in this rock is as follows. Consider that the rock contains 5500 molecules of MgO and 4500 molecules of SiO_2. Now, using the Mg_2SiO_4 formula, ol contains 2 molecules of MgO and 1 molecule of SiO_2; while using $MgSiO_3$, opx contains 1 molecule of MgO and 1 molecule of SiO_2. If there are a molecules of Mg_2SiO_4 and b molecules of $MgSiO_3$, then:

$$2a + b = 5500$$
$$a + b = 4500$$

which solves to give $a = 1000$ and $b = 3500$; or:

$$\text{mole proportion ol} = \frac{1000}{1000 + 3500} = 0·2222$$

$$\text{mole proportion opx} = \frac{3500}{1000 + 3500} = 0·7778$$

Note that if the opx had been written on double this formula unit as $Mg_2Si_2O_6$, then:

$$2a + 2b = 5500$$
$$a + 2b = 4500$$

and $a = 1000$, $b = 1750$; giving:

mole proportion of ol $= 0.3636$

mole proportion of opx $= 0.6364$

Thus the mole proportions are dependent on the formulae of the phases involved in the calculation.

The lever rule is an equivalent way of performing this calculation. This rock composition would plot between FO and EN on figure 2.9(a). Now using the lever rule:

$$\text{mole proportion ol} \quad = \frac{0.55 - 0.5}{0.6666 - 0.5} = 0.3$$

$$\text{mole proportion opx} = \frac{0.6666 - 0.55}{0.6666 - 0.5} = 0.7$$

Now, why are these numbers different to the ones calculated above? The reason comes from the fact that $MgSiO_3$, $Mg_2Si_2O_6$ etc. all plot on the same position in figure 2.9(a); whereas, as shown above, the formulae chosen for the phases affect the calculated mole proportions. The lever rule actually gives the mole proportions of formulae whose MgO and SiO_2 molecules sum to 1. For olivine this formula is $Mg_{\frac{2}{3}}Si_{\frac{1}{3}}O_{\frac{4}{3}}$ and for orthopyroxene, $Mg_{\frac{1}{2}}Si_{\frac{1}{2}}O_{\frac{3}{2}}$. These mole proportions can be converted to the mole proportions of preferred formulae by dividing the mole proportions by the appropriate total number of MgO and SiO_2 molecules in the preferred formulae, and resumming to 1. Thus for

		Mole proportions
Mg_2SiO_4	$\dfrac{0.3}{3} = 0.1$	$\dfrac{0.1}{0.45} = 0.2222$
$MgSiO_3$	$\dfrac{0.7}{2} = 0.35$	$\dfrac{0.35}{0.45} = 0.7778$
	0.45	

and for:

		Mole proportions
Mg_2SiO_4	$\dfrac{0.3}{3} = 0.1$	$\dfrac{0.1}{0.275} = 0.3636$
$Mg_2Si_2O_6$	$\dfrac{0.7}{4} = 0.175$	$\dfrac{0.35}{0.275} = 0.6364$
	0.275	

To calculate the weight proportions, we need the molecular weight of the olivine $(2(40\cdot32)+60\cdot09 = 140\cdot73)$ and the orthopyroxene $(40\cdot32 + 60\cdot09 = 100\cdot41)$. Using MP for mole proportions:

Mineral	MP	MP×MW	Weight proportion
olivine	0·2222	31·270	$\dfrac{31\cdot270}{109\cdot369}=0\cdot2859$
orthopyroxene	0·7778	78·099	$\dfrac{78\cdot099}{109\cdot369}=0\cdot7141$
		109·369	

Note that the weight proportions of the minerals are independent of the formulae chosen for the phases, because, for example, if $Mg_2Si_2O_6$ is chosen for the formula of the orthopyroxene, then the different mole proportions are counteracted by the molecular weight of the orthopyroxene now being 200·82.

To calculate the volume proportions, we need the volume (V) of Mg_2SiO_4 $(4\cdot379$ kJ kbar$^{-1})$ and $MgSiO_3$ $(3\cdot144$ kJ kbar$^{-1})$. Thus:

Mineral	MP	MP× V	Volume proportion
olivine	0·2222	0·9729	0·2846
orthopyroxene	0·7778	2·4454	0·7154
		3·4183	

Note that the volume and weight proportions are very similar in this case. This is because the densities of the two phases are very similar (32·14 for olivine, 31·94 for orthopyroxene, both in g kJ^{-1} kbar).

2(c) Plot the positions of talc (TA, $Mg_3Si_4O_{10}(OH)_2$), and chrysotile (CH, $Mg_3Si_2O_5(OH)_4$) on a triangular composition diagram representing the ternary system MgO–SiO_2–H_2O.

The lever rule principle also operates in triangular phase diagrams. A phase plotting at a on figure 2.9(b) would be the fictitious phase $Mg_3O_2(OH)_2$, a phase consisting of just H_2O and MgO in the proportion $\frac{1}{4}$ to $\frac{3}{4}$, or in the ratio 1:3. All the compositions along the line joining a and the SiO_2 apex of the triangle will also have the ratio of 1:3 for $H_2O:MgO$. Thus talc will plot somewhere along this line. Talc must also lie somewhere along the line from b to MgO, on which compositions will have the ratio of 1:4 for $H_2O:SiO_2$. Thus talc plots at the intersection of these two lines. The same procedure can be repeated for chrysotile.

Triangular diagrams are used as compatibility diagrams for ternary (three end-member) systems in phase diagrams. Figure 2.9(c) shows part of a compatibility diagram which might be appropriate to the formation conditions of serpentinites. A line connecting two phases, called a tie line, indicates that these two phases can coexist in stable equilibrium for appropriate rock compositions at the conditions to which the compatability diagram applies. A rock composition will contain those three phases which define the apices of the triangle within which the rock composition plots. Which triangles are stable within the compatibility diagram can be determined from the relative positions of the $G-x$ loops for the phases in a $G-x$ diagram which has G as the vertical axis from the triangular diagram as its base.

Any rock plotting within the shaded area on figure 2.9(c) will contain chrysotile–talc–fluid. The proportions of the phases depending on where the rock composition plots within the shaded area. The actual proportions are obtained using the reverse of the procedure for plotting compositions in the $MgO-SiO_2-H_2O$ diagram. For rock composition A, the ratio of the proportions of TA and CH, TA/CH, is $3 \times \frac{7}{8} = 2 \cdot 625$, by applying the lever rule to the line TA–CH where A–F (projected) intersects it, noting that CH contains 7 and talc contains 8 molecules of MgO, SiO_2 and H_2O. Similarly $F/CH = 0 \cdot 1 \times \frac{7}{1} = 0 \cdot 7$. As we know that the proportions must add up to one, then $TA + CH + F = 1$. Substituting the above relations into this:

$$CH + 2 \cdot 625CH + 0 \cdot 7CH = 1$$

Solving, $CH = 0 \cdot 231$, and substituting back, $TA = 0 \cdot 607$ and $F = 0 \cdot 162$.

From a compatibility diagram point of view, we need not always plot assemblages in the system $MgO-SiO_2-H_2O$ on a triangular diagram. Consider a sequence of metamorphic rocks which had an aqueous fluid present during metamorphism. One apex of all the compatibility triangles applicable to these assemblages will always be F. Therefore all the compatibility information that we require can be plotted on a $MgO-SiO_2$ line by projecting the minerals and rocks which plot within the triangle from H_2O onto the $MgO-SiO_2$ side of the triangle, figure 2.9(d). Rock compositions can be plotted on this binary representation to give the assemblage, with the assumption that every assemblage is 'plus F'. Figure 2.9(e) is equivalent to figure 2.9(c) showing the rock composition A plotting in the stability field of TA–CH–F. The proportions of the phases obtained from such a projection are the proportions of the projected phases. Figure 2.9(e) will give the proportions of the solid phases in the assemblage, $TA = 0 \cdot 724$ and $CH = 0 \cdot 276$. Other useful projections can be devised, depending on the mineralogical characteristics of the rocks of interest. For example, there is a well known projection which is very

useful for pelitic rocks which involves projecting from quartz and muscovite as well as H_2O in the system $MgO–FeO–K_2O–Al_2O_3–SiO_2–H_2O$.

2(d)   Construct the $P–T$ diagram for the aluminosilicate (Al_2SiO_5) system involving sillimanite (S), andalusite (A) and kyanite (K) from
the $G–T$ diagrams (figure 2.10) assuming that the reaction lines on

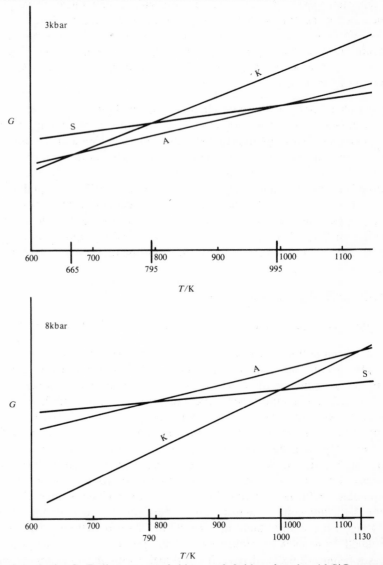

2.10   Schematic $G–T$ diagrams at 3 kbar and 8 kbar for the Al_2SiO_5 system
involving andalusite (A), kyanite (K) and sillimanite (S), from which the
aluminosilicate phase diagram can be constructed. (WE 2(d))

the *P–T* diagram are straight. At what pressure and temperature
can S, A and K coexist in stable equilibrium?

The first step is to transfer the temperature of the intersections of the
G–T lines for each mineral pair at each pressure onto a *P–T* diagram
(figure 2.11) using solid symbols for stable intersections and open symbols
for metastable intersections. In figure 2.10(a), the intersection of the
sillimanite and kyanite lines takes place at a higher Gibbs energy than
andalusite, so the reaction involving sillimanite and kyanite is metastable

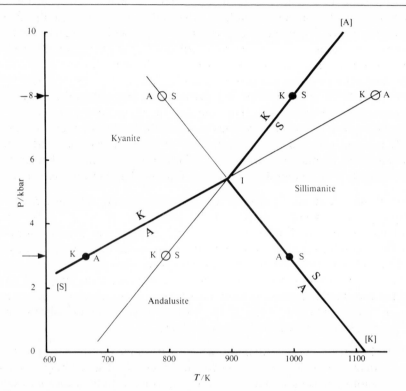

2.11 The aluminosilicate *P–T* phase diagram constructed from the *G–T* diagrams in figure 2.10. The triple point, where kyanite + andalusite + sillimanite can
be stable together, is labelled I. (WE 2(d))

with respect to andalusite at these conditions. Each point on the *P–T*
diagram can be labelled on the high and low temperature sides with the
phase which is relatively more stable on that side of the point.

The appropriate pairs of points can now be joined up. The next step is
to differentiate between the stable and metastable parts of these reaction
lines. The segments of lines through the open symbols are metastable,
while the segments through the closed symbols are stable. Each line

changes from stable to metastable across the common intersection of the three lines, the so called triple point. This can be seen to be correct by envisaging the transformation of figure 2.10(a) to 2.10(b) by changing pressure. This involves moving the andalusite line relative to the other lines. At some pressure between 3 and 8 kbar, the andalusite line will pass through the intersection of the kyanite and sillimanite lines. This intersection corresponds to the triple point I on the *P–T* diagram, where andalusite, sillimanite and kyanite can all coexist in stable equilibrium.

Each of the *P–T* fields between stable lines on figure 2.11 can now be labelled with the stable assemblage or assemblages, in this case the stable aluminosilicate. Each stable line can be labelled with the reaction which takes place across it. It can also be labelled with the phase *not* involved in the reaction, thus the A = K reaction is the sillimanite-out reaction (meaning the reaction not involving sillimanite) written as [S]. Note that the metastable extension of [S] occurs between sillimanite-producing reactions (K = S and A = S). The rule that the metastable extension of [*i*] lies between *i*-producing reactions is a general feature of these diagrams and is often referred to as Schreinemaker's rule. It can be used in the construction of diagrams when less information is available than in this problem. It is also helpful in checking a finished diagram for consistency.

Andalusite, sillimanite and kyanite can coexist at stable equilibrium at the triple point at about 900 K and 5·5 kbar. The presence of all three polymorphs in a rock does not necessarily indicate that the rock formed at these conditions. This is because the aluminosilicates, once formed in an assemblage, react away only sluggishly when they become metastable. The presence of all three aluminosilicates may indicate that the *P–T* path of the rock during metamorphism passed through the stability fields of the three aluminosilicates, each of the earlier formed aluminosilicates not reacting away as a new one becomes stable.

2(e) Construct the *T–x* diagram which corresponds to the series of *G–x* diagrams in figure 2.12. Give the sequence of assemblages found with increasing temperature for rock compositions *a* and *b* assuming equilibrium at each temperature, giving also the approximate proportions of the minerals in the assemblages, and the way the compositions of the minerals change with temperature.

The first step is to draw the compatibility diagram for each of the *G–x* diagrams. The range of solution of each of the phases is small because the *G–x* loops are tightly curved. The compatibility diagrams can now be assembled on a *T–x* diagram (figure 2.13). Filling in the *T–x* diagram can be achieved by visualizing how each *G–x* diagram is transformed into the next one with increasing temperature. For example, from figure 2.12(a) to 2.12(b) the *G–x* loop for C has crossed the common tangent involving A and D. Thus at some temperature between these two diagrams the *G–x*

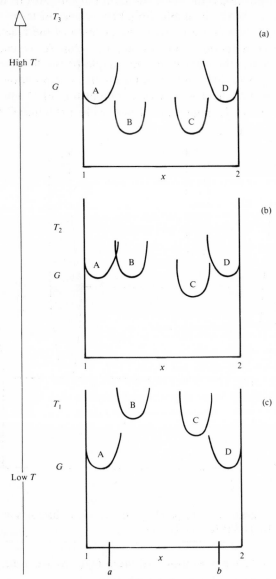

2.12 $G-x$ diagrams for three temperatures (at equal temperature intervals) for a binary system between end-members 1 and 2 involving phases A, B, C and D, all of which show only a limited range of solution between the end-members. (WE 2(e))

loop for C will just touch the A–D common tangent. At this temperature, A, C and D can coexist in stable equilibrium, and, in terms of changing temperature, this temperature marks the position of the reaction A + D = C. On the $T–x$ diagram a horizontal line joins A, C and D at this temperature. Note the shapes of the single-phase fields on the $T–x$ diagram, particularly that the C field terminates at a point on the A–C–D line. This is because there is only one composition of C which can be in equilibrium with A + D, regardless of how wide a solution C can form, in

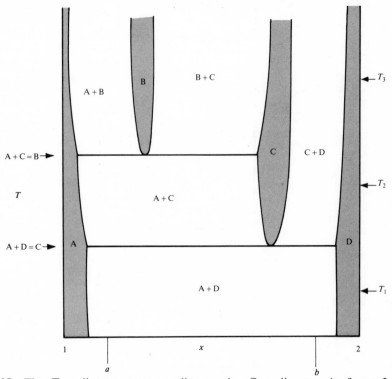

2.13 The $T–x$ diagram corresponding to the $G–x$ diagrams in figure 2.12. One-phase fields are shaded. (WE 2(e))

this case, at higher temperatures. All these features are a direct consequence of the geometry of the $G–x$ diagrams.

For rock composition a, the sequence of assemblages with increasing temperature is A + D, A + C, A + B. In the A + D assemblage, the (mole) proportions are approximately 0·08 for D and 0·92 for A. In the A + C assemblage, the proportions are changing with temperature because the compositions of A and C are changing with temperature, A and C both becoming richer in end-member 1. Across the A + C zone, the proportions change from 0·11 of C and 0·89 of A to 0·16 of C and 0·84 of A with increasing temperature. In the A+B zone, the proportions also change with increasing temperature, from 0·44 of B and 0·56 of A at the appearance of B with increasing temperature to 0·62 of B and 0·38 of A at the top of the diagram. The way the proportions of the phases change with increasing temperature are summarized in figure 2.14(a).

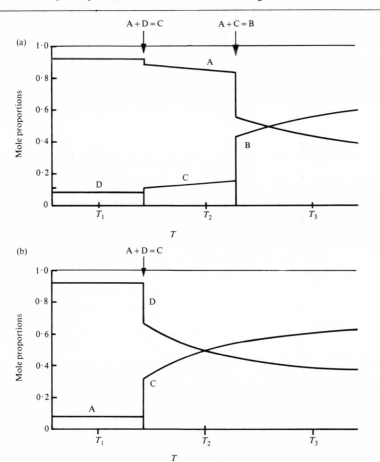

2.14 The mole proportions of the phases in compositions *a* and *b* in figure 2.13 as a function of temperature, (a) and (b) respectively. (WE 2(e))

The sequence of assemblages for rock composition *b* is A+D then C+D with increasing temperature. The way the proportions of the phases change with increasing temperature are summarized in figure 2.14(b). Note that the reaction A+C=B does not affect this rock composition.

2(f) Answer question 1(e) for figure 2.15.

As in the answer to 1(e), the first step is to draw the compatibility diagram for each of the *G–x* diagrams. These can then be assembled on a

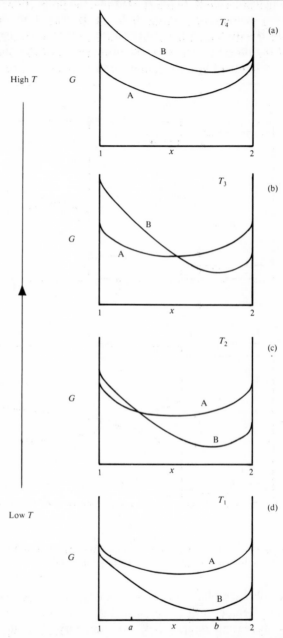

2.15 $G-x$ diagrams for four temperatures (at equal temperature intervals) for a binary system between end-members 1 and 2 involving phases A and B, both of which can show complete solid solution between the end-members. (WE 2(f))

T–x diagram (figure 2.16). As before, filling in the diagram involves visualizing the transformation of one G–x diagram into the next with changing temperature. Here one G–x loop is moving across the other with changing temperature, and the effect on the T–x diagram is to describe a T–x loop.

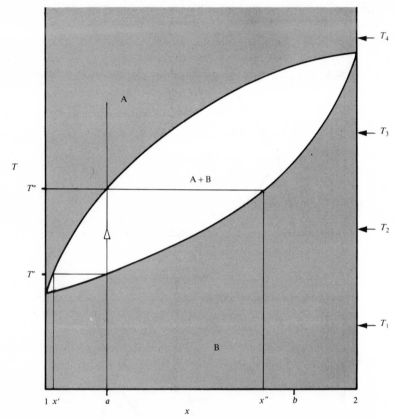

2.16 The T–x diagram corresponding to the G–x diagrams in figure 2.15. One-phase fields are shaded. (WE 2(f))

Consider rock composition a. At low temperatures a consists of just phase B, which therefore also has the composition of a. With increasing temperature, the T–x loop is intersected at T'. At this point, phase B is joined by an infinitesimal amount of phase A composition x'. As temperature is increased further the compositions and proportions of A and B change until at T'', phase A of composition a is in equilibrium with an infinitesimal amount of B of composition x''. Above T'', phase A of composition a is the stable assemblage in the rock. The same logic can be applied to rock composition b. The proportions of the phases in rock compositions a and b as a function of temperature are summarized in figure 2.17.

2(g) Show the different types of binary T–x diagrams that are generated by the intersection of an open G–x loop for a higher temperature phase C with the G–x loops for two lower temperature phases A and B, if (i) A and B have the same structure, and (ii) A and B have different structures; and for different curvatures of the loops for A and B.

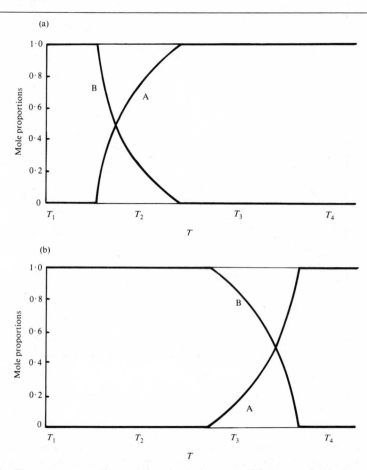

2.17 The mole proportions of the phases in compositions a and b in figure 2.16 as a function of temperature, (a) and (b) respectively. (WE 2(f))

The G–x loops for phases A and B will be continuous if the phases have the same structure (figures 2.18(a)–(c)). The result of decreasing the curvature of the G–x loops for the two phases will be to stabilize a complete solution between the two phases, denoted by A′ (figure

2.18(c)). Clearly this cannot happen if the two phases have a different structure. The composition gap between two phases of the same structure is usually called a solvus, while the composition gap between phases of different structure is usually called a miscibility gap.

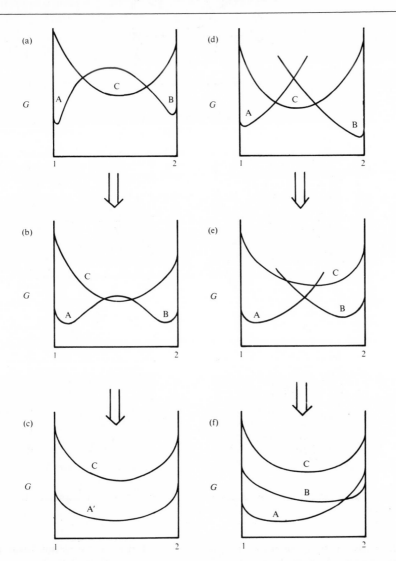

2.18 Two series of G–x diagrams ((a)–(c), (d)–(f)), each with decreasing 'dissimilarity' between the end-members for phases A and B: (a)–(c) for phases A and B having the same structure; (d)–(f) for A and B having different structures. Phase C can show a complete range of solution between the end-members 1 and 2. The relative position of the G–x loop for C moves downwards with respect to the G–x loops for A and B with increasing temperature for each diagram. (WE 2(g))

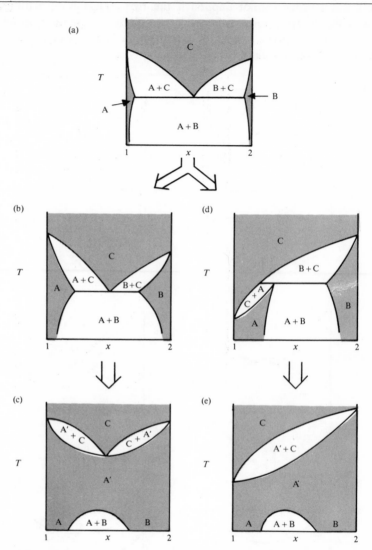

2.19 The different types of $T-x$ phase diagram which correspond to the $G-x$ diagrams in figure 2.18(a)–(c). (WE 2(g))

The possible types of phase diagram if A and B have the same structure are shown in figure 2.19. There are two main variations depending on whether the $G-x$ loop for C intersects the common tangent between A and B before it intersects the $G-x$ loop for A (consisting of pure 1), (b) and (c), or after it intersects the $G-x$ loop for A, (d) and (e), both with increasing temperature. The solvus between A and B in (c) and (e) may be at too low a temperature to be observable.

Some of the possible types of phase diagrams if A and B have different structures are shown in figure 2.20. Note that figure 2.19(a) is the same as figure 2.20(a), although the miscibility gap between A and B in figure 2.20(a) is not a solvus with a truncated top but a much fatter version of

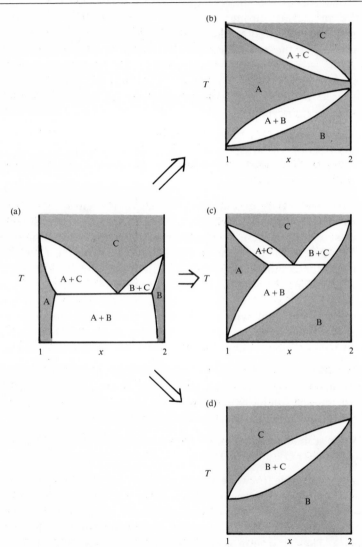

2.20   Some of the different types of T–x phase diagram which correspond to the G–x diagrams in figure 2.18(d)–(f). (WE 2(g))

the T–x loop between A and B in figure 2.20(c). Which of (b)–(d) is applicable in figure 2.20 depends on the relative positions of the three open G–x loops for A, B and C. In (d) phase A is metastable. An alternative set of diagrams can be generated if the $A+C$ loop in figure 2.20(a) terminates at the 2- rather than the 1-axis. This will result in the $A+C$ loops in (b) and (c) also sloping the other way.

These diagrams are applicable to the melting of silicates in binary systems, with C as the silicate liquid, and can be used to illustrate the sequence of events during melting or crystallization. Frequently for diagrams like figure 2.20(a), the range of mutual solubility of the minerals is so small that the one-phase field is less than the width of the line marking the axis, and so is not represented. The diagrams are also useful for metamorphic systems, particularly diagrams like figure 2.20(c), and can be used to show how the compositions of minerals change with progressive metamorphism.

2(h) Show the different binary $T-x$ diagrams that are generated by the intersection of an open $G-x$ loop for a phase D with the tightly curved $G-x$ loops for phases A, B and C, all of which have different structures.

The three main types of diagram are shown in figure 2.21. In (a) B becomes metastable with respect to $A+C$ before any composition of D becomes stable, with increasing temperature. In this case the upper part of the $T-x$ diagram is the same as figure 2.20(a). In (b), the $G-x$ loop for D intersects the common tangents between A and B and between B and C before engulfing the loop for B, with increasing temperature. The result is two diagrams like figure 2.20(a) placed side by side. In (c), this is not true because B breaks down to $D+C$ before B can break down directly to D as in (b). If these are considered as melting diagrams (D is the liquid), B is said to melt congruently in (b) and to melt incongruently in (c).

As more phases are added to the binary system, the number of possible types of diagram increases, particularly if some of the phases show extended mutual solubility. Knowing the possible relationships on $T-x$ diagrams is important in interpreting assemblages in rocks.

2(i) Write a balanced reaction between talc (TA), chrysotile (CH), forsterite (FO), and H_2O (F).

We need to be able to write balanced chemical reactions because the chemical potential expression for a reaction is the starting point for thermodynamic calculations. We are particularly interested in reactions between the end-members of minerals, rather than between the minerals themselves. The latter approach is more useful, for example, for relating the assemblages either side of an isograd in a metamorphic terrain by the isograd reaction.

This reaction, like many that we need to write, can be written down by inspection. However, here, we will go through the general method to see how it works. First, we see what a reaction looks like on composition diagrams. Consider figures 2.22(a) and (b). These show how compatibility diagrams are simply projections onto the base of the prism of the triangles between G–x loops which define minimum Gibbs energy for compositions within the triangle. In figure 2.22(a), for composition A, the EN–CH–TA assemblage has a lower Gibbs energy than EN–CH–F; while in figure 2.22(b), the opposite is true. The difference, then, between the two diagrams is that in (a), the tie line between TA and CH is stable,

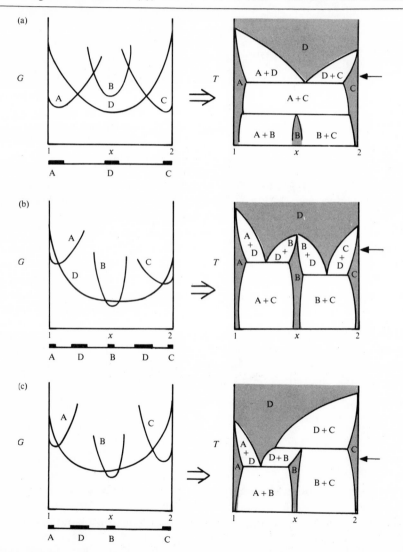

2.21 Three G–x and the corresponding T–x phase diagrams for the intersection of an open loop for a phase D, and three tightly curved loops for phases A, B and C. (WE 2(h))

whereas in (b), the tie line between EN and F is stable. These two compatibility diagrams are related by the reaction:

$$TA + CH = EN + F$$

From the point of view of assemblages, this does not mean that as this reaction is crossed that TA and CH completely disappear, to be replaced

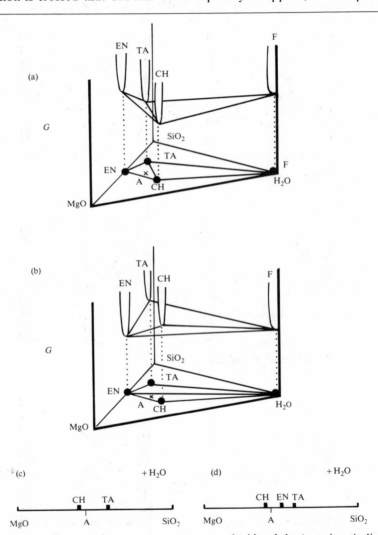

2.22 G–x–x diagrams for two temperatures each side of the 'crossing tie lines' reaction $TA + CH = EN + H_2O$, showing the relationship between the G–x–x planes between the phases and the tie lines on the compatibility diagrams which form the base of each prism. (c) and (d) are these compatibility diagrams presented as projections from H_2O onto MgO–SiO_2. The mineralogy for rock composition A is discussed in the text.

by EN and F. It does mean that assemblages involving TA *plus* CH are not stable when the EN–F tie line becomes stable. When the reaction is actually taking place, the G–x loops of TA, CH, EN, and F are coplanar, they lie on the same plane. When this occurs, tie lines between CH and TA and between EN and F are both stable. This reaction is therefore characterized by crossing tie lines. Note that crossing tie lines will only occur on a compatibility diagram if the compatibility diagram refers to conditions at which the reaction is taking place.

Projections of compatibility diagrams are just as useful for considering reactions. Figures 2.22(c) and (d) show the compatibility diagrams for this reaction after projecting from F. Note that these compatibility diagrams do not indicate which side of the reaction F occurs on.

The diagrammatic approach will reveal how many phases are required to write a balanced reaction. For example, in a triangular diagram, four are generally required, but in certain cases three are sufficient. In figure 2.9(b), as Q, EN, and FO are collinear, they lie on a straight line, a reaction can be written between just these three phases. In using projected compatibility diagrams, the number of phases required to write a reaction must include the phase (or phases) which are projected from, if they are needed to balance the reaction. This can only be seen after finding the reaction coefficients for the other phases.

The calculation of reaction coefficients involves solving a set of simultaneous equations (see appendix B, for a method of solving simultaneous equations). The first step is to set up the simultaneous equations. The unknowns are the reaction coefficients, which will be denoted by their respective names. A phase with a positive reaction coefficient occurs on the right-hand side of the reaction, while a phase with a negative reaction coefficient occurs on the left-hand side of the reaction. We can write an equation for each oxide which must be obeyed if the reaction is balanced with respect to that oxide. Thus for MgO:

$$3(TA) + 3(CH) + 1(EN) + 0(F) = 0$$

The coefficients in this equation are the number of MgO molecules in the formula unit of the phase. Similarly, for SiO_2 and H_2O:

$$4(TA) + 2(CH) + 1(EN) + 0(F) = 0$$
$$1(TA) + 2(CH) + 0(EN) + 1(F) = 0$$

The equations cannot be solved yet as there are only three equations for the four unknowns. However consider the two equations:

$$Mg_2SiO_4 + SiO_2 = 2MgSiO_3$$
$$3Mg_2SiO_4 + 3SiO_2 = 6MgSiO_3$$

Both of these are the same balanced reaction, only in one all the reaction coefficients are three times larger than in the other. Thus, we have to fix the reaction coefficient of one phase in the reaction before we can calculate the other reaction coefficients. Solving the set of equations is simplified if the chosen reaction coefficient is for a phase which occurs in several equations. In this case, setting $TA = 1$, the equations become:

(a) $3CH + EN$ $= -3$

(b) $2CH + EN$ $= -4$

(c) $2CH$ $+ F = -1$

Subtracting (b) from (a) gives $CH = 1$. Substituting this value into (a) gives $EN = -6$. Substituting these values into (c) gives $F = -3$. The equation can now be written out:

$$6MgSiO_3 + 3H_2O = Mg_3Si_4O_{10}(OH)_2 + Mg_3Si_2O_5(OH)_4$$
    enstatite   fluid   talc   chrysotile

It is always worthwhile checking to see if the reaction actually does balance for each oxide, as a check on the arithmetic. If any of the reaction coefficients had come out as fractions, then all the reaction coefficients would be multiplied by a factor to make all the reaction coefficients whole numbers.

When projections are being used for compatibility diagrams, the writing of balanced reactions can be simplified. Consider that H_2O is considered to be always present in this system. The simultaneous equations are then set up ignoring the fluid phase and the equation for balancing H_2O. H_2O is balanced after the other reaction coefficients have been calculated. So if this case, with $TA = 1$:

$3CH + EN = -3$

$2CH + EN = -4$

Then $CH = 1$ and $EN = -6$. The incomplete reaction is then written:

(incomplete) $$6MgSiO_3 = Mg_3Si_4O_{10}(OH)_2 + Mg_3Si_2O_5(OH)_4$$
    enstatite   talc   chrysotile

The number of H_2O molecules can then be obtained by inspection, giving the required reaction as above.

2(j) Write the equilibrium relation for the assemblage enstatite (EN), quartz (Q), chrysotile (CH) and talc (TA).

In figure 2.23(a), for composition A, the EN–TA–CH assemblage has a lower Gibbs energy than EN–Q–CH, while in figure 2.23(b) the opposite is true. The difference between the two diagrams is the stability of TA. These two compatibility diagrams are related by the reaction:

$TA = EN + CH + Q$

This reaction is characterized by the disappearance of a phase (TA). Reactions are either of this type or of the crossing tie lines type. Note that the reaction relating figures 2.22(c) and (d) is of the disappearance of phase type, even though the unprojected compatibility diagrams (figures 2.22(a) and (b)) are related by a crossing tie lines reaction.

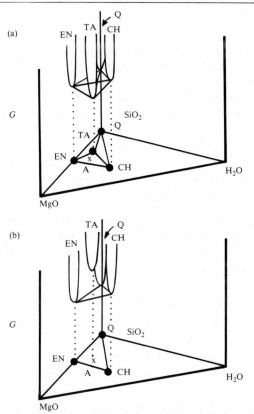

2.23 G–x–x diagrams for two temperatures either side of the 'disappearance of phase' reaction TA = EN + Q + CH, showing the relationship between the G–x–x planes between the phases and the tie lines on the compatibility diagrams which form the base of each prism. (WE 2(j))

In the calculation of reaction coefficients, take TA = 1, then:

for MgO $3CH + EN$ $= -3$

for SiO_2 $2CH + EN + Q = -4$

for H_2O $2CH$ $= -1$

Thus, $CH = -\frac{1}{2}$, $EN = -\frac{3}{2}$ and $Q = -\frac{3}{2}$. Multiplying all the reaction coefficients by two gives:

$$3SiO_2 + 3MgSiO_3 + Mg_3Si_2O_5(OH)_4 = 2Mg_3Si_4O_{10}(OH)_2$$
quartz   enstatite   chrysotile   talc

Thus, the equilibrium relation is:

$$3\mu_{SiO_2,quartz} + 3\mu_{MgSiO_3,enstatite} + \mu_{Mg_3Si_2O_5(OH)_4,chrysotile} = 2\mu_{Mg_3Si_4O_{10}(OH)_2,talc}$$

2(k) How many reactions can be written between the phases calcite (CC), dolomite (DOL), talc (TA), tremolite (TR) assuming a CO_2–H_2O fluid phase is always present? Are all these reactions independent?

This system can be plotted on a triangular diagram, as a result of a projection from H_2O and CO_2, with CaO, MgO and SiO_2 as apices (figure 2.24(a)). As no three phases are collinear in the diagram, all reactions will involve four phases. Each reaction involves all the phases bar one. Therefore, the number of reactions is five. The reaction can be labelled by the omitted phase as in worked example 2(d). The reactions are:

[Q] $3CC + 2TA = DOL + TR + CO_2 + H_2O$
[CC] $4Q + 2DOL + TA = TR + 4CO_2$
[DOL] $4Q + 6CC + 5TA = 3TR + 6CO_2 + 2H_2O$
[TA] $8Q + 5DOL + H_2O = 3CC + TR + 7CO_2$
[TR] $4Q + 3DOL + H_2O = 3CC + TA + 3CO_2$

The equilibrium relations for these are obtained as usual by rewriting each in terms of chemical potentials. All these equilibrium relations would apply to the assemblage $Q + CC + DOL + TA + TR$ if it formed at equilibrium. However not all are independent. Subtracting [CC] from [Q] we obtain:

$$3CC + TA + 3CO_2 = 3DOL + 4Q + H_2O$$

which is none other than [TR]. Similarly, [TA] is obtained by subtracting twice [CC] from [Q]; [DOL] is obtained by adding twice [Q] to [CC]. Thus, just two of the equilibrium relations are independent for this assemblage, meaning that the other equilibrium relations can be derived from these two.

2(l) How many equilibrium relations can be written between the phases talc (TA), tremolite (TR), diopside (DI), calcite (CC) and quartz (Q) assuming a CO_2–H_2O fluid phase always present?

These can be plotted on a CaO–MgO–SiO_2 diagram as for the last worked example (figure 2.24(b)). Note that here TA, TR, and DI are collinear. This means that [Q] and [CC] are the same reaction involving TA, TR and DI. Such a reaction involving less than the usual number of phases is called degenerate. There are four reactions giving four equilibrium relations, with only two being independent as before. The reactions are:

[DI] $4Q + 6CC + 5TA = 3TR + 6CO_2 + 2H_2O$
[Q, CC] $TR = TA + 2DI$
[TR] $3CC + 2Q + TA = 3DI + 3CO_2 + H_2O$
[TA] $TR + 3CC + 2Q = 5DI + 3CO_2 + H_2O$

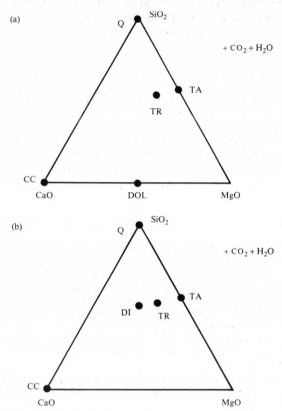

2.24   The positions of calcite (CC), dolomite (DOL), tremolite (TR), quartz (Q) and talc (TA) on a CaO–MgO–SiO$_2$ diagram which is a projection from H$_2$O and CO$_2$. (WE 2(k))

Problems 2

2(a)   A rock consists of 50 mole % magnetite (Fe$_3$O$_4$) and 50 mole % quartz (SiO$_2$). What are the weight and volume proportions of the magnetite and quartz in the rock? (The molecular weights and volumes of magnetite and quartz are 231·55 and 60·09, and 4·452 and 2·269 respectively.)

2(b)   Plot enstatite (EN), MgSiO$_3$; forsterite (FO), Mg$_2$SiO$_4$; kyanite (KY), Al$_2$SiO$_5$; spinel (SP), MgAl$_2$O$_4$; and pyrope (PY), Mg$_3$Al$_2$-Si$_3$O$_{12}$ on a MgO–SiO$_2$–Al$_2$O$_3$ triangular diagram.

2(c)   What are the mole proportions of PY and KY in a rock which plots half way between PY and KY in the answer to 2(b)?

2(d)   Write all the reactions between the phases in 2(b).

2(e)   Plot grossular (GR), $Ca_3Al_2Si_3O_{12}$; quartz (Q), SiO_2; zoisite (Z), $Ca_2Al_3Si_3O_{12}(OH)$; anorthite (AN), $CaAl_2Si_2O_8$; and sillimanite (S), Al_2SiO_5 on a $CaO–SiO_2–Al_2O_3$ triangular with H_2O treated as being in excess.

2(f)   Plot the diagram for 2(e) with quartz in excess (as well as H_2O).

2(g)   Write all the reactions between the phases in 2(e) for: (a) H_2O in excess; (b) $Q+H_2O$ in excess.

2(h)   Find the reaction involving anorthite ($CaAl_2Si_2O_8$); phlogopite ($KMg_3AlSi_3O_{10}(OH)_2$); tremolite ($Ca_2Mg_5Si_8O_{22}(OH)_2$); orthoclase ($KAlSi_3O_8$); zoisite ($Ca_2Al_3Si_3O_{12}(OH)$); chlorite (Mg_5Al_2-$Si_3O_{10}(OH)_8$) and H_2O. (Use the method in appendix B.)

References

Smith, E. B., 1973, *Basic Chemical Thermodynamics*, Oxford University Press, London: This is an excellent introductory text for students who have little or no physical chemistry background. For these students, this book should be parallel reading for chapters 2–6.

Denbigh, K., 1971, *An Introduction to Chemical Equilibrium,* Cambridge University Press, Cambridge: This is a good orthodox chemical thermodynamics book, which can be treated as advanced reading on the chemical aspects of this book. The development is considerably more mathematical than the development in Smith's book.

Broecker, W. S., and Oversby, V. M., 1971, *Chemical Equilibria in the Earth*, McGraw-Hill, New York: A geology text covering much of the same ground as chapters 1–6, but with a more mathematical approach.

Ehlers, E.G., 1972, *The Interpretation of Geological Phase Diagrams*, Freeman, San Francisco: Chapters 1–3 are recommended parallel reading for students with little or no previous exposure to phase diagrams.

Chapter 3

Activities and Standard states

Each chemical potential is a function of temperature, pressure and composition because the Gibbs energy from which the chemical potential is derived is a function of these parameters. For convenience, the chemical potential is split into two terms, a composition-independent term, called the standard chemical potential, and a term which is a function of composition which accounts for the difference between the actual chemical potential for the composition of the phase of interest and the standard chemical potential. Thus:

$$\mu_{1A} = \mu_{1A}^{\circ} + RT \ln a_{1A} \tag{3.1}$$

where μ_{1A}° is the standard chemical potential of end-member 1 in phase A, a_{1A} is the activity of end-member 1 in phase A, R is the gas constant ($R = 0.0083144 \text{ kJ K}^{-1}$) and T is the temperature in K ($= {}^{\circ}\text{C} + 273$). The standard state, to which the standard chemical potential refers, is some temperature, pressure, and composition of that phase. The form of the second term, the composition-dependent term, is convenient for expressing the composition dependence of the chemical potential of end-members in many phases—as we shall see later. The activity, a_{1A}, is a function of the composition of phase A, and also often temperature and pressure. When the actual phase is in the standard state, the activity must equal 1, so that the logarithm of the activity is equal to 0.

Standard state 1

The obvious choice of standard state (standard state 1) is the pure end-member at the temperature and pressure of interest. This means that the standard chemical potential refers to the pure end-member at the pressure and temperature of interest, or:

$$\mu_{1A}^{\circ} = \mu_{1A(x_{1A}=1)} = G_{A(x_{1A}=1)} \equiv G_{1A}$$

where G_{1A} is the Gibbs energy of phase A when it consists of just end-member 1. The activity term takes account of the difference in composition between the actual phase being considered and the pure end-member. If the actual phase consists of just the pure end-member 1, then $a_{1A} = 1$ (and $RT \ln a_{1A} = 0$), as the phase corresponds to the standard state conditions. The activity is a more or less complicated function of composition (and temperature and pressure), and an important part of the application of equilibrium thermodynamics is the consideration of activity–composition $(a-x)$ relations, i.e. the consideration of the dependence of activity on composition (and temperature and pressure). Fortunately, under some conditions, the $a-x$ relations are simple. This can be seen with the help of figure 3.1, a diagram of chemical potential of component 1 in a binary phase A. There are three main regions.

Raoult's law region

This region extends from pure end-member 1, $x_{1A} = 1$ (i.e. $\ln x_{1A} = 0$), to compositions where end-member 1 is diluted somewhat by end-member 2. The region is characterized by a straight line segment on a μ_{1A} against $\ln x_{1A}$ (μ–$\ln x$) plot (a–b in figure 3.1), of slope RT. As μ_{1A} for pure 1 is G_{1A}, then the equation of this line is:

$$\mu_{1A} = G_{1A} + RT \ln x_{1A} \tag{3.2}$$

For our standard state 1, $\mu_{1A}^{\circ} = G_{1A}$, and so, by comparison with (3.1), this region is characterized by $a_{1A} = x_{1A}$.

Henry's law region

This region extends from infinite dilution of end-member 1 in end-member 2, or $x_{1A} \rightarrow 0$, meaning that the phase consists predominantly of 2 with very little of 1. This region is also characterized by a straight line segment on a μ–$\ln x$ plot (c–d in figure 3.1) of slope RT. The intercept on the μ_{1A}-axis is not now G_{1A} but G_{1A} plus a term. This term is a function of the end-member which end-member 1 is infinitely diluted by, temperature and pressure, but not x_{1A}. This term can be expressed in the form $RT \ln h_{1A(2)}$, where $h_{1A(2)}$ is the Henry's law constant for end-member 1 in phase A infinitely diluted by end-member 2. Thus the equation for this line is:

$$\mu_{1A} = G_{1A} + RT \ln x_{1A} + RT \ln h_{1A(2)}$$

Combining the two logarithmic terms:

$$\mu_{1A} = G_{1A} + RT \ln x_{1A} h_{1A(2)} \tag{3.3}$$

For our standard state 1, $\mu_{1A}^\circ = G_{1A}$, and so by comparison with (3.1), this region is characterized by $a_{1A} = x_{1A} h_{1A(2)}$.

Intermediate region

This region lies between the above two regions. It is characterized by a curved line segment on a μ–ln x plot (b–c in figure 3.1) which connects

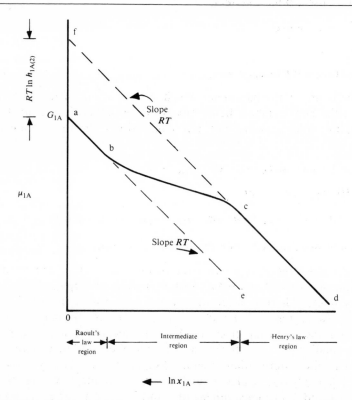

3.1   A plot of μ_{1A} against ln x_{1A} showing the form of the composition dependence of the chemical potential; with three regions, the Raoult's law, intermediate and Henry's law regions.

the two straight line segments. In terms of an equation for the chemical potential, this region is characterized by an equation which is intermediate between (3.2) and (3.3). Introducing a new function, the activity coefficient, γ_{1A}, we can write:

$$\mu_{1A} = \mu_{1A}^\circ + RT \ln x_{1A} \gamma_{1A}$$

with γ_{1A} a function of x_{1A} in such a way that:

in the Raoult's law region, $\gamma_{1A} = 1$,   so that $a_{1A} = x_{1A}$

in the Henry's law region, $\gamma_{1A} = h_{1A(2)}$, so that $a_{1A} = x_{1A} h_{1A(2)}$

Many different equations for the activity coefficients as a function of composition have been devised, all obeying the above constraints for the Henry's law and Raoult's law regions. These equations are called mixing models because they reflect the interactions between (the atoms of) the end-members mixing in the phase. One such model is the regular model. This will be used here to demonstrate some features of $a-x$ relations. For this model for a binary system:

$$RT \ln \gamma_{1A} = w x_{2A}^2$$

so that:

$$\mu_{1A} = G_{1A} + RT \ln x_{1A} + w x_{2A}^2 \tag{3.4}$$

where w is an energy parameter which is a reflection of the 'difference' between the end-members being mixed in the phase—it is a function of temperature and pressure but not composition. In the Raoult's law region:

$$x_{1A} \to 1 \quad \text{so} \quad w x_{2A}^2 \to 0 \quad \text{or} \quad \gamma_{1A} \to 1$$

In the Henry's law region:

$$x_{1A} \to 0 \quad \text{so} \quad w x_{2A}^2 \to w \quad \text{or} \quad \gamma_{1A} = h_{1A(2)} \to \exp(w/RT)$$

This model does indeed obey the above constraints.

Figure 3.1 is a general representation of the composition dependence of the chemical potential in a binary system (and, in principle, more complicated systems). There is a strong correlation between the width of the Raoult's law region and the size of $h_{1A(2)}$. The more similar are end-members 1 and 2, the larger are the Raoult's law and Henry's law regions and the smaller is $h_{1A(2)}$. If the two end-members are very similar then $h_{1A(2)} \to 1$, the intermediate region disappears, the Henry's law and Raoult's law regions coalesce, and we refer to ideal mixing of end-members 1 and 2. For ideal mixing of 1 and 2, $a_{1A} = x_{1A}$ (i.e. $\gamma_{1A} = 1$) and $a_{2A} = x_{2A}$ (i.e. $\gamma_{2A} = 1$). From the point of view of performing calculations involving chemical potentials, the assumption of ideal mixing for a phase is often used. The assumption of ideal mixing is equivalent to using line a–e in figure 3.1 for the chemical potential. The incorrectness of this is given by the distance between a–e and a–b–c–d for a particular composition, x_{1A}. Thus the assumption will be correct in the Raoult's law region, but it will become progressively more incorrect in moving across the intermediate region. The amount of this incorrectness in the Henry's law region depends on the size of the Henry's law constant (or $RT \ln h_{1A(2)}$ in figure 3.1). Figure 3.2 shows these features on a series of μ–ln x diagrams which have been calculated using the regular model with different values of w.

In summary, the chemical potential of end-member 1 in phase A is given by:

$$\mu_{1A} = G_{1A} + RT \ln x_{1A} \gamma_{1A} \qquad (3.5)$$

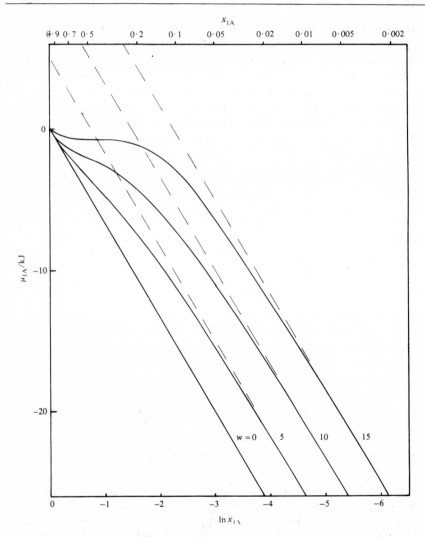

3.2 A plot of μ_{1A} against ln x_{1A} for the regular solution model using values of w of 0(5)15 ($w = 0$ corresponding to an ideal solution); showing the decreasing size of the Raoult's law and Henry's law regions as w increases.

where $\gamma_{1A} \to 1$ when $x_{1A} \to 1$ and $\gamma_{1A} \to$ constant when $x_{1A} \to 0$, and this is most simply expressed with standard state 1, with:

$$\mu_{1A}^{\circ} = G_{1A} \quad \text{and} \quad a_{1A} = x_{1A}\gamma_{1A}$$

This is the most commonly used standard state, mainly for end-members of solids, but also for certain end-members of fluids.

Standard state 2

Standard state 1 is only useful if it is possible to make phase A consisting of pure 1. If we are considering a rare-earth element, say lanthanum, La, in a pyroxene, it is not possible to make a pure La pyroxene, so that G_{1A} cannot be determined directly. Similarly, if we are considering NaCl species in an aqueous solution, we cannot make pure NaCl in the water structure. The situation is portrayed in figure 3.3. Here, μ_{1A} can be measured for small values of x_{1A} (large negative values of ln x_{1A}), but we

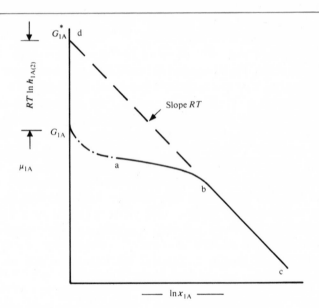

3.3 A plot of μ_{1A} against ln x_{1A} illustrating standard state 2.

know nothing about it as $x_{1A} \to 1$ (the chain line). The extension of the Henry's law straight line segment b–c intersects the μ_{1A}-axis at $x_{1A} = 1$ (or ln $x_{1A} = 0$) at d. The Gibbs energy here can be referred to as G_{1A}^{*}. Although $G_{1A}^{*} = G_{1A} + RT \ln h_{1A(2)}$, we do not know G_{1A} and $RT \times \ln h_{1A(2)}$ individually. But for the Henry's law region:

$$\mu_{1A} = G_{1A}^{*} + RT \ln x_{1A} \tag{3.6}$$

Departures from the Henry's law region into the intermediate region can be accounted for by an activity coefficient, γ_{1A}^*, whose form must be governed by the constraint that $\gamma_{1A}^* = 1$ in the Henry's law region. Thus:

$$\mu_{1A} = G_{1A}^* + RT \ln x_{1A} \gamma_{1A}^* \tag{3.7}$$

Using the regular model, knowing that $RT \ln h_{1A(2)} = w$:

$$G_{1A}^* = G_{1A} + w$$

Substituting this in (3.4):

$$\mu_{1A} = G_{1A}^* - w + RT \ln x_{1A} + w(1 - x_{1A})^2$$
$$= G_{1A}^* + RT \ln x_{1A} + w x_{1A}(x_{1A} - 2)$$

Or, for this mixing model:

$$RT \ln \gamma_{1A}^* = w x_{1A}(x_{1A} - 2)$$

In general, the relationship between the activity coefficients is:

$$RT \ln \gamma_{1A}^* = RT \ln \gamma_{1A} - RT \ln h_{1A(2)}$$

A different standard state will be more convenient here. This standard state (standard state 2) is a hypothetical pure end-member whose properties are obtained by extrapolating from the Henry's law region at the $P-T$ of interest, thus $\mu_{1A}^\circ = G_{1A}^*$ and $a_{1A} = x_{1A}\gamma_{1A}^*$. An important qualification about the usefulness of this standard state arises because $G_{1A}^* = G_{1A} + RT \ln h_{1A(2)}$. The value of G_{1A}^* depends on the end-member in which end-member 1 is infinitely diluted. Thus, while $G_{\mathrm{MgSiO_3,cpx}}$ is independent of the composition of the clinopyroxene, $G_{\mathrm{NiSiO_3,cpx}}^*$ could be strongly dependent on the composition of the clinopyroxene. For this reason this standard state is not very useful in trace element geochemistry because the host phase for the trace element may show a wide range of solid solution. However for phases which are dominated by one end-member this standard state is useful. A standard state similar to this is used for dissolved species in aqueous solutions. This will be covered later.

Standard state 3

Sometimes pure end-member 1 in the structure of phase A cannot be easily studied whereas pure end-member 1 in the structure of B can. Consider figure 3.4. Now, as usual, we have:

$$\mu_{1A} = G_{1A} + RT \ln x_{1A} \gamma_{1A}$$

Adding $G_{1B} - G_{1B}$ to the right-hand side of this equation:

$$\mu_{1A} = G_{1B} + (G_{1A} - G_{1B}) + RT \ln x_{1A} \gamma_{1A}$$

Putting the bracketed term inside the logarithmic term:

$$\mu_{1A} = G_{1B} + RT \ln\left[x_{1A}\gamma_{1A}\exp\left(\frac{G_{1A}-G_{1B}}{RT}\right)\right]$$

or:

$$\mu_{1A} = G_{1B} + RT \ln x_{1A}\gamma'_{1A} \quad \text{where} \quad \gamma'_{1A} = \gamma_{1A}\exp\left(\frac{G_{1A}-G_{1B}}{RT}\right) \quad (3.8)$$

This amounts to a new standard state (standard state 3), which is defined as the pure end-member in a different structure to the phase at the

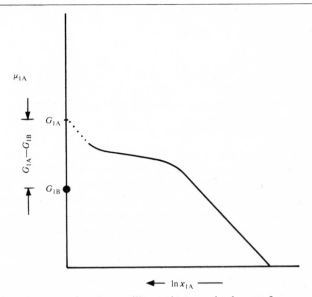

3.4   A plot of μ_{1A} against $\ln x_{1A}$ illustrating standard state 3.

temperature and pressure of interest. Thus $\mu^\circ_{1A} = G_{1B}$ and $a_{1A} = x_{1A}\gamma'_{1A}$. Note that as $x_{1A} \to 1$, $a_{1A} \to \exp[(G_{1A}-G_{1B})/RT]$. This standard state has been used for end-members in silicate liquids with the standard chemical potential referring to the Gibbs energy of the pure end-member as a solid or glass, rather than as a liquid.

Standard state 4

The previous standard states have all referred to the temperature and pressure of interest. Under certain circumstances, it is useful to have a standard state of a pressure of 1 bar and the temperature of interest. Consider the effect of this variant on the equations of standard state 1, starting with:

$$\mu_{1A} = G_{1A} + RT \ln x_{1A}\gamma_{1A}$$

Splitting the G_{1A} term (see (A.1) in appendix A) and moving the volume integral into the logarithmic term:

$$\mu_{1A} = G_{1A}(1, T) + \int_1^P V_{1A}dP + RT \ln x_{1A}\gamma_{1A}$$

$$= G_{1A}(1, T) + RT \ln\left[x_{1A}\gamma_{1A} \exp\left(\frac{\int_1^P V_{1A}dP}{RT}\right)\right] \tag{3.9}$$

Noting that V_{1A} refers to the volume of pure end-member 1 at the P–T of interest, whereas x_{1A} and γ_{1A} refer to the actual composition of the gas at the P–T of interest. This, (3.9), amounts to a statement of a standard state (standard state 4) of pure end-member at 1 bar and the temperature of interest, with:

$$\mu_{1A}^\circ = G_{1A}(1, T) \quad \text{and} \quad a_{1A} = x_{1A}\gamma_{1A} \exp\left(\frac{\int_1^P V_{1A}\,dP}{RT}\right)$$

noting that the activity equals the exponential term for pure end-member 1. This standard state is used for some end-members of fluids.

Summary

In the preceding sections, the idea of a standard state has been developed with the aid of four examples. The important point is that although a standard state is an arbitrary device for splitting the chemical potential into composition-dependent, $RT \ln a_{1A}$, and composition-independent, μ_{1A}, terms, we do know the form of the chemical potential,

$$\mu_{1A}^\circ = G_{1A} + RT \ln x_{1A}\gamma_{1A} \tag{3.10}$$

with $\gamma_{1A} \to 1$, when $x_{1A} \to 1$ and $\gamma_{1A} \to$ constant, when $x_{1A} \to 0$. All standard states can be derived from this, as it stands or by appropriate re-arrangement. It is most important to realize that a particular choice of standard state for an end-member means using a particular μ_{1A}° *and the* particular a_{1A}.

We can now return to the equilibrium relation for a balanced chemical reaction, $\Delta\mu = 0$. For example consider the metamorphic limestone assemblage, quartz (Q), dolomite (DOL), diopside (DI) and fluid (F). This can be considered with the following reaction:

$$CaMg(CO_3)_2 + 2SiO_2 = CaMgSi_2O_6 + 2CO_2$$
$$\quad \text{DOL} \qquad\quad \text{Q} \qquad\qquad \text{DI} \qquad\quad \text{F}$$

for which the equilibrium relation is:

$$\mu_{CaMg(CO_3)_2,DOL} + 2\mu_{SiO_2,Q} = \mu_{CaMgSi_2O_6,DI} + 2\mu_{CO_2,F}$$

Each of these chemical potentials can be expressed in terms of standard chemical potentials and activities, thus:

$$\mu^{\circ}_{CaMg(CO_3)_2,DOL} + 2\mu^{\circ}_{SiO_2,Q} + RT \ln a_{CaMg(CO_3)_2,DOL} + 2RT \ln a_{SiO_2,Q}$$
$$= \mu^{\circ}_{CaMgSi_2O_6,DI} + 2\mu^{\circ}_{CO_2,F} + RT \ln a_{CaMgSi_2O_6,DI} + 2RT \ln a_{CO_2,F} \quad (3.11)$$

Using ΔG° to mean the change of the standard chemical potentials for the reaction, then:

$$\Delta G^{\circ} = \mu^{\circ}_{CaMgSi_2O_6,DI} + 2\mu^{\circ}_{CO_2,F} - \mu^{\circ}_{CaMg(CO_3)_2,DOL} - 2\mu^{\circ}_{SiO_2,Q} \quad (3.12)$$

Noting that this is (products) − (reactants) for the reaction, ΔG° is referred to as the standard Gibbs energy of reaction, meaning the change in Gibbs energy for the reaction with each of the end-members in its standard state. Similarly the activity terms can be summed. Using $RT \ln K$ to mean the change of the $RT \ln a$ terms for the reaction, then:

$$RT \ln K = RT \ln a_{CaMgSi_2O_6,DI} + 2RT \ln a_{CO_2,F} - RT \ln a_{CaMg(CO_3)_2,DOL}$$
$$- 2RT \ln a_{SiO_2,Q}$$

Noting that $y \ln x = \ln x^y$, this can be simplified to:

$$RT \ln K = RT \ln \frac{a_{CaMgSi_2O_6,DI} a^2_{CO_2,F}}{a_{CaMg(CO_3)_2,DOL} a^2_{SiO_2,Q}} \quad (3.13)$$

Noting that this is (products)/(reactants), K is called the equilibrium constant. Comparing (3.11) to (3.13), the equilibrium relation for a reaction can be expressed as:

$$\Delta\mu = 0 = \Delta G^{\circ} + RT \ln K \quad \text{or} \quad -\Delta G^{\circ} = RT \ln K \quad (3.14)$$

This expression is quite general and is simply a restatement of $\Delta\mu = 0$ using a standard state for each end-member. Its simplicity should not allow one to forget that for each end-member, the chosen standard state means that each standard chemical potential is associated with a particular form for the activity.

Worked examples 3

3(a) Draw the G–x diagram and compatibility diagram for a binary system at $T = 800$ K which is: (a) an ideal solution; (b) a regular solution with $w = 7$ kJ; (c) a regular solution with $w = 18$ kJ.

A new equation is needed for this calculation. For a binary phase A:

$$G_A = x_{1A}\mu_{1A} + x_{2A}\mu_{2A} = x_{1A}G_{1A} + x_{2A}G_{2A} + x_{1A}RT \ln x_{1A}\gamma_{1A}$$
$$+ x_{2A}RT \ln x_{2A}\gamma_{2A}$$

For these G–x diagrams we will take $G_{1A} = G_{2A} = 0$; this does not affect the compatibility diagrams which will be generated. Thus, substituting this and the regular model equations (3.4) into this equation and noting that $x_{2A} = 1 - x_{1A}$:

$$G_A = x_{1A}RT \ln x_{1A} + (1 - x_{1A})RT \ln (1 - x_{1A}) + x_{1A}w(1 - x_{1A})^2$$
$$+ (1 - x_{1A})wx_{1A}^2$$

Simplifying:

$$G_A = RT[x_{1A} \ln x_{1A} + (1 - x_{1A}) \ln (1 - x_{1A})] + wx_{1A}(1 - x_{1A})$$

The $G-x$ diagrams in figure 3.5 were plotted using this equation, using $w = 0$ for (a). Note that as the non-ideality becomes stronger (i.e. as w increases), the depth of the $G-x$ loop decreases, until in (c), the single loop now consists of two bumps. This means that the compatibility diagram involves two phases, rather than the complete solution present at lower w values. Actually, there will be two phases in the compatibility diagram if w is greater than $2R$ times the temperature of interest. Thus, at 800 K, for w greater than $2(0 \cdot 0083144)800 = 13 \cdot 3$ kJ, there will be two phases in the compatibility diagram.

3(b) Draw $G-x$ diagrams for a binary system which is a regular solution with $w = 18$ kJ at 700, 800, 900, 1000, 1100 and 1200 K in order to plot a $T-x$ diagram. Outline the crystallization behaviour of a system with composition $x_1 = 0 \cdot 2$ cooling from 1100 K.

The $G-x$ curves are plotted in figure 3.6 (a). The line A–B gives the locus of the compositions of the coexisting phases which defines a solvus. This information can also be presented on a $T-x$ diagram (figure 3.6 (b)). The temperature of the top of the solvus, the consolute temperature, T_c, can be found using the formula in the answer to the last worked example:

$$T_c = \frac{w}{2R} = \frac{18}{2(0 \cdot 0083144)} = 1080 \text{ K}$$

The regular model can only generate symmetrical solvi. Two or more mixing parameters are required to generate an asymmetric solvus in a binary system (see worked example 3(d)).

At 1100 K, a system of composition $x_1 = 0 \cdot 2$ consists of just one phase with the same composition as the system. On cooling, at 940 K, the system intersects the solvus. The phase exsolves an infinitesimal amount of a phase of the same structure but of composition $x_1 = 0 \cdot 8$. With further cooling the composition of the host becomes more 1-rich, while the exsolved phase becomes more 2-rich, by further exsolution. Exsolution usually takes the form of lamellae and blebs, for example in the alkali feldspars and pyroxenes. Progressive exsolution with cooling is indicated in, for example, pyroxenes, by several sets of exsolution lamellae, which themselves may contain exsolution lamellae. Exsolution in most silicates does not continue much below 500 °C even for very slow cooling rates

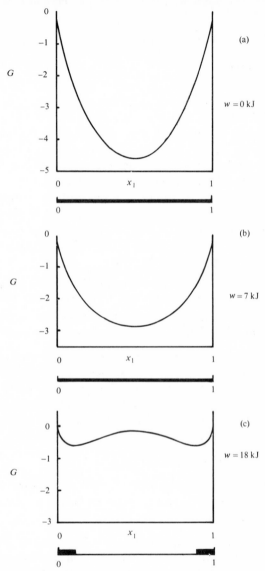

3.5 G–x diagrams for different values of w at a temperature of 800 K for the regular solution model, showing the appearance of a solvus as w increases at a particular temperature. (WE 3(a))

because the atoms cannot diffuse far enough or fast enough to continue forming exsolution lamellae, i.e. equilibrium is not maintained during cooling below this temperature. The amount of exsolution and the temperature down to which exsolution continues to occur depends on the cooling history of the mineral. If cooling is rapid then exsolution may not occur. Exsolution can also occur in systems with $T-x$ diagrams like figure 2.16, particularly if the difference in temperature between the top and bottom of the $T-x$ loop is very large. The pyroxene $T-x$ diagrams are composite examples of this type.

3(c) Show that the mixing model for a binary system in which the

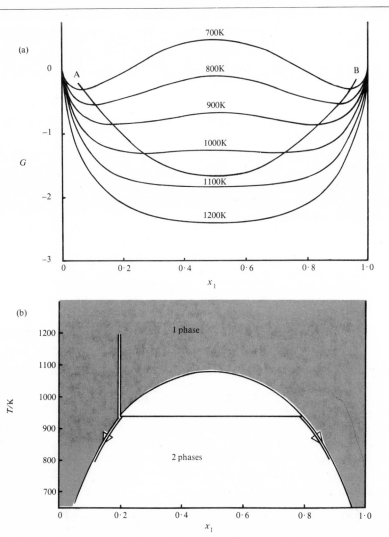

3.6 The $G-x$ diagram, (a), shows $G-x$ curves for a series of temperatures for the regular solution model with $w = 18$ kJ. The curve A-B is the locus of the compositions of the coexisting phases below the top of the solvus. The $T-x$ diagram, (b), shows this solvus, generated by decreasing temperature at a particular value of w. (WE 3(b))

activity coefficients are given by:

$$RT \ln \gamma_{1A} = \frac{A_1}{\left(1 + \dfrac{A_1 x_{1A}}{A_2 x_{2A}}\right)^2} \qquad RT \ln \gamma_{2A} = \frac{A_2}{\left(1 + \dfrac{A_2 x_{2A}}{A_1 x_{1A}}\right)^2}$$

obeys Raoult's law and Henry's law.

This mixing model is called the van Laar model. It involves two parameters, A_1 and A_2 in a binary system, rather than one as in the regular model. If $A_1 = A_2$, and noting that $x_{2A} = 1 - x_{1A}$, then:

$$RT \ln \gamma_{1A} = A_1(1 - x_{1A})^2 \quad \text{and} \quad RT \ln \gamma_{2A} = A_1(1 - x_{2A})^2$$

Thus the van Laar model reduces to the regular model when $A_1 = A_2$.

In the Raoult's law region for end-member 1, $x_{1A} \to 1$ and $x_{2A} \to 0$. Therefore x_{1A}/x_{2A} is much larger than 1, so $RT \ln \gamma_{1A} \to 0$. In the Henry's law region for end-member 1, $x_{1A} \to 0$ and $x_{2A} \to 1$. Therefore x_{1A}/x_{2A} is much less than 1, so $RT \ln \gamma_{1A} \to A_1$. In the Henry's law region for end-member 2, $RT \ln \gamma_{2A} \to A_2$. Thus, this mixing model is asymmetric as the Henry's law constants for the two end-members are different, in contrast to the regular model.

3(d)   Draw G–x diagrams for a binary system which is a van Laar solution with $A_1 = 15 \, \text{kJ}$ and $A_2 = 20 \, \text{kJ}$ at 700, 800, 900, 1000, 1100 and 1200 K in order to plot a T–x diagram.

The G–x diagrams in figure 3.7(a) were calculated using:

$$G_A = x_{1A} RT \ln x_{1A} + x_{2A} RT \ln x_{2A} + \frac{x_{1A} A_1}{\left(1 + \dfrac{A_1 x_{1A}}{A_2 x_{2A}}\right)^2} + \frac{x_{2A} A_2}{\left(1 + \dfrac{A_2 x_{2A}}{A_1 x_{1A}}\right)^2}$$

Simplifying:

$$G_A = RT[x_{1A} \ln x_{1A} + (1 - x_{1A}) \ln (1 - x_{1A})] + \frac{x_{1A}(1 - x_{1A})A_1 A_2}{A_2(1 - x_{1A}) + A_1 x_{1A}}$$

The corresponding T–x diagram is given in figure 3.7(b). Note that the solvus is asymmetric. The solvus becomes more asymmetric as A_1/A_2 departs further from 1. The steeper limb of the solvus occurs on the side with the larger A term associated with it. In this case A_2 is larger so the steeper limb is the x_2-rich limb.

3(e)   Expand the chemical potentials in the equilibrium expression for the reaction involving enstatite, talc, chrysotile and fluid using standard state 1 for each end-member.

The reaction is:

$$Mg_3Si_4O_{10}(OH)_2 + Mg_3Si_2O_5(OH)_4 = 6MgSiO_3 + 3H_2O$$

     talc   chrysotile   enstatite   fluid

For the equilibrium coexistence of talc, chrysotile, enstatite and fluid:

$$\Delta\mu = 0 = \Delta G° + RT \ln K$$

where:

$$\Delta G° = 6G_{MgSiO_3,\text{enstatite}} + 3G_{H_2O,\text{fluid}} - G_{Mg_3Si_4O_{10}(OH)_2,\text{talc}}$$

$$- G_{Mg_3Si_2O_5(OH)_4,\text{chrysotile}}$$

An expression for K is given below. The expression for $\Delta G°$ is incomplete because the question as set is not precise enough. The expression

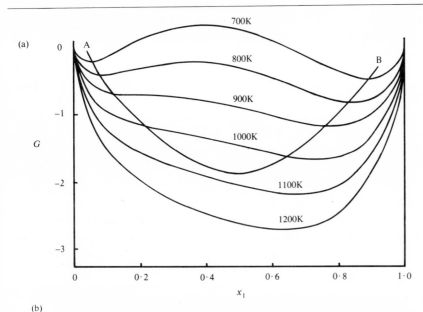

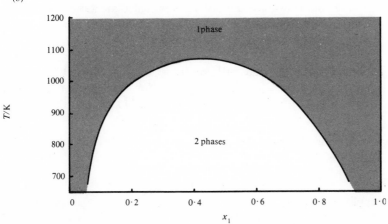

3.7 As for figure 3.6, but using the van Laar model with $A_1 = 15$ kJ and $A_2 = 20$ kJ, generating an asymmetric solvus. (WE 3(d))

for $\Delta G°$ contains the Gibbs energy of $MgSiO_3$ in enstatite. But enstatite occurs in several different polymorphs, orthoenstatite, clinoenstatite and protoenstatite. The Gibbs energy of $MgSiO_3$ in each of these structures is different, and thus the result of any calculation using the equilibrium relation will be wrong if the Gibbs energy for the wrong polymorph is used. Thus, when the reaction is written down and in the subscript to the Gibbs energies, it should be clear which structure the end-member of interest is assumed to be in. For example, a more precise statement of the reaction is:

$$Mg_3Si_4O_{10}(OH)_2 + Mg_3Si_2O_5(OH)_4 = \quad 6MgSiO_3 \quad + 3H_2O$$
$$\text{talc} \qquad\qquad \text{clinochrysotile} \qquad \text{orthoenstatite} \qquad \text{fluid}$$

for which the equilibrium relation is:

$$\Delta\mu = 0 = \Delta G° + RT \ln K$$

where:

$$\Delta G° = 6G_{MgSiO_3,\text{orthoenstatite}} + 3G_{H_2O,\text{fluid}}$$
$$- G_{Mg_3Si_4O_{10}(OH)_2,\text{talc}} - G_{Mg_3Si_2O_5(OH)_4,\text{clinochrysotile}}$$

and:

$$K = \frac{a^6_{MgSiO_3,\text{orthoenstatite}}\, a^3_{H_2O,\text{fluid}}}{a_{Mg_3Si_4O_{10}(OH)_2,\text{talc}}\, a_{Mg_3Si_2O_5(OH)_4,\text{clinochrysotile}}}$$

where $a_i = x_i\gamma_i$. Such expressions may seem unnecessarily cumbersome, but ambiguities can easily result in‑using shorthand like a_{TA} for the activity of $Mg_3Si_4O_{10}(OH)_4$ in talc, particularly when considering complicated solid solutions which can occur in several structures. Of course, abbreviations for the end-member and for the phase in which the end-member is located can remove some of the long-windedness.

3(f) Using the answer to the last section, calculate the dependence on fluid composition of the temperature of formation of the assemblage, orthoenstatite, talc, clinochrysotile and fluid at 2 kbar, when the talc consists of just $Mg_3Si_4O_{10}(OH)_2$; clinochrysotile of just $Mg_3Si_2O_5(OH)_4$, and orthoenstatite of just $MgSiO_3$; given that $\Delta G° = 238\cdot8 - 0\cdot261T$ kJ at 2 kbar (for the reaction TA+CH = 6EN+3H_2O), and that the fluid is an ideal solution (i.e. $a_{H_2O,\text{fluid}} = x_{H_2O,\text{fluid}}$).

The equilibrium relation for the assemblage at 2 kbar for the above stipulations is:

$$\Delta\mu = 0 = 238\cdot8 - 0\cdot261T + RT \ln x^3_{H_2O,F}$$

where T is in K and F stands for fluid. Substituting $R = 0\cdot0083144$ kJ K^{-1} and simplifying:

$$T = \frac{9570}{10\cdot5 - \ln x_{H_2O}} \quad \text{(in K)}$$

A new sort of T–x diagram, of much use in metamorphic petrology, can be used here. This is a plot of temperature against composition of the fluid phase (in this case x_{H_2O}) on which the positions of reactions can be plotted (figure 3.8). The labelling of the reaction (i.e. which is the low temperature side of the reaction and which is the high temperature side) can be performed using the minimization of Gibbs energy principle. For $x_{H_2O} = 1$, the difference in Gibbs energy between the products and the reactants is $\Delta\mu = 238\cdot8 - 0\cdot261T$. Now, at 800 K, a temperature below the calculated position for the reaction at $x_{H_2O} = 1$, $\Delta\mu = 30\cdot0$ kJ. As $\Delta\mu$ is positive at this temperature it means that the Gibbs energy of the products is greater than the Gibbs energy of the reactants. This, in turn,

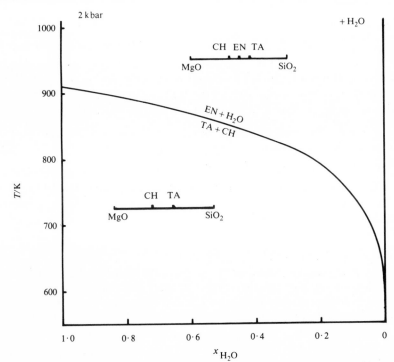

3.8   The T–x diagram for the reaction TA + CH = EN + H$_2$O at 2 kbar. (WE 3(f))

means that the reactants form the stable tie line at low temperature. As a rule of thumb the fluid comes out on the high temperature side of the reaction. The reaction can now be labelled. The fields on the diagram can also be labelled with compatibility diagrams (see for example figure 2.22).

The dependence of the temperature of the reaction on the fluid composition becomes progressively stronger as the H$_2$O in the fluid phase becomes more and more diluted (by for example CO$_2$). The reaction only reaches $x_{H_2O} = 0$ at $T = 0$ K which means that the reaction is asymptotic to the axis at $x_{H_2O} = 0$. At another pressure, the reaction will have the same shape but will have moved upwards or downwards with respect to the line in figure 3.8, depending on the pressure term in the equilibrium relation. The reaction forms a surface on a P–T–x diagram.

3(g) What are the shapes of the lines for reactions involving H_2O and/or CO_2 in T–x_{CO_2} diagrams?

A general reaction involving H_2O and/or CO_2 can be written in terms of n H_2O and m CO_2 molecules, where n can be negative, zero or positive and m can be negative, zero or positive. Writing the $\Delta G°$ for each reaction as $\Delta G°/R = a - bT$ at a particular pressure, then, for pure solid phases and ideal mixing of H_2O and CO_2, re-arranging (3.14) gives:

$$T = \frac{a}{b - n \ln x_{H_2O} - m \ln x_{CO_2}}$$

The different shapes can be considered by substituting different values of n and m into this equation. The values substituted are on the assumption that the fluid end-members are on the right-hand, high temperature, side of the reaction, so that a and b are both always positive.

(i) $n = m = 0$. The reaction is a solid–solid reaction, $T = a/b$, and is represented by a horizontal line on a T–x diagram because the temperature of the reaction does not depend on x_{CO_2} (figure 3.9, line (i)).

(ii) $n = 1$, $m = 0$. The reaction is a dehydration reaction. Here:

$$T = \frac{a}{b - \ln x_{H_2O}}$$

As x_{H_2O} has to be less than or equal to 1, $\ln x_{H_2O}$ is less than or equal to 0. Thus T decreases progressively as x_{H_2O} becomes smaller (line (ii) in figure 3.9).

(iii) $n = 0$, $m = 1$. The reaction is a decarbonation reaction. Here:

$$T = \frac{a}{b - \ln x_{CO_2}}$$

T decreases progressively as x_{CO_2} becomes smaller (line (iii) in figure 3.9).

(iv) $n = 1$, $m = 1$. The reaction evolves both CO_2 and H_2O. Here:

$$T = \frac{a}{b - \ln x_{H_2O} - \ln x_{CO_2}}$$

T decreases progressively as x_{CO_2} becomes small and as x_{H_2O} becomes small. Thus the line has a maximum. The fluid composition for this maximum is given by $dT/dx_{CO_2} = 0$. Performing this differentiation, using $x_{H_2O} = 1 - x_{CO_2}$, n and m for the numbers of molecules of H_2O and CO_2

in the reaction, and x for x_{CO_2} at the maximum:

$$\frac{dT}{dx_{CO_2}} = \frac{-a\left(-\dfrac{n}{x}+\dfrac{m}{1-x}\right)}{b-n\ln x-m\ln(1-x)} = 0$$

Simplifying:

$$x = \frac{n}{n+m}$$

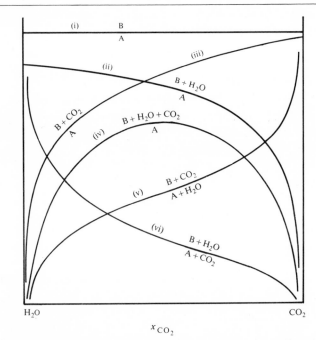

3.9 The shape of the lines for the five different types of reaction on a T–x_{CO_2} diagram. (**WE** 3(g))

In this case the composition of the fluid for the maximum temperature is $x_{CO_2} = 0.5$ (line (iv) in figure 3.9). If the reaction involved $n = 3$ and $m = 1$, then the composition of the fluid at the maximum is $x_{CO_2} = 0.75$, and so on.

(v) $n = 2$, $m = -1$. The reaction involves H_2O and CO_2 on opposite sides of the reaction. CO_2 will be consumed on the low temperature side of the reaction and H_2O will be evolved on the high temperature side because the magnitude of n is larger than the magnitude of m. Here:

$$T = \frac{a}{b-2\ln x_{H_2O}+\ln x_{CO_2}}$$

In this case T decreases progressively as x_{H_2O} becomes small, while T increases progressively as x_{CO_2} becomes small (line (v) in figure 3.9). The temperature goes to infinity when $b = 2 \ln x_{H_2O} - \ln x_{CO_2}$. For normal values of b, this occurs very close to $x_{CO_2} = 0$, as drawn in figure 3.9. However, for a small value of b, the temperature goes to infinity at a larger value of x_{CO_2}.

If n is greater than 0 and m is less than 0, and $n + m = 0$ then the slope of the T–x line has to be obtained by calculating the position of the line from thermodynamic data as no generalization can be made about the sign of a and b.

(vi) $n = -1$, $m = 2$. H_2O is consumed on the low temperature side of the reaction and CO_2 evolved on the high temperature side. Here:

$$T = \frac{a}{b + \ln x_{H_2O} - 2 \ln x_{CO_2}}$$

T decreases progressively as x_{CO_2} becomes small, while T increases progressively as x_{H_2O} becomes small (line (vi) in figure 3.9). The size of b has the same relevance here as in (v) above.

Problems 3

3(a) Show that the mixing model for a binary system in which the activity coefficients are given by:

$$RT \ln \gamma_{1A} = x_{2A}^2 [w_1 + 2x_{1A}(w_2 - w_1)]$$
$$RT \ln \gamma_{2A} = x_{1A}^2 [w_2 + 2x_{2A}(w_1 - w_2)] \quad \text{(subregular model)}$$

(a) obeys Raoult's law and Henry's law; (b) reduces to the regular model if $w_1 = w_2$.

3(b) Determine the shape of the solvus in a binary system which is a subregular solution with $w_1 = 15$ kJ and $w_2 = 20$ kJ from G–x diagrams at 100 K intervals from 700 K to 1000 K. ($R = 0.0083144$ kJ K^{-1}.)

3(c) Write the equilibrium relations for the reactions in problem 2(g).

References

Smith, E. B., 1973, *Basic Chemical Thermodynamics*, Oxford University Press, London: Introductory text, see notes on p 44.

Denbigh, K., 1971, *An Introduction to Chemical Equilibrium*, Cambridge University Press, Cambridge: An advanced text; see notes on p 44.

Powell, R., 1974, A comparison of some mixing models for crystalline silicate solid solutions. *Contr. Mineral. Petrol.*, **46**, 265–74: Advanced reading on mixing models, including regular, subregular and van Laar models.

Chapter 4

Thermodynamics of Solids

The Gibbs energy of a phase, and thus the chemical potentials of its end-members, is dependent on its state of aggregation. The term solid can refer to crystalline matter which has a well defined three-dimensional lattice produced by the regular arrangement of atoms, each atom being in a state of thermal motion about an average position on the lattice. The term solid can also be applied to amorphous solids which have no large-scale *regular* lattice. However there can be every transition between a large single crystal and an amorphous solid in terms of the size of individual volumes which have a lattice in a particular orientation. In a large single crystal, the whole crystal consists of a lattice in a particular orientation, while an amorphous solid consists of very many small volumes each having a regular lattice, these lattices being randomly orientated within the solid. The 'grain size', either of individual crystals in a polycrystalline aggregate (for example a clay) or of each volume of a solid which has a regular lattice, only affects the Gibbs energy of the grains via a term for the surface energy, which is proportional to the surface area of the grains. The surface area of a mole of phase increases dramatically as the grain size becomes very small. The surface energy term only becomes significant when the grain size approaches clay particle size.

The rest of this section will be addressed to the properties of crystalline solids with a relatively large-scale regular lattice. However, no crystal is without imperfections (defects) which affect this regular lattice. These can take many forms, from vacancies and interstitial atoms (zero-dimensional defects), dislocations (one-dimensional defects) to stacking faults (two-dimensional defects). Fortunately under normal circumstances these defects have a negligible effect on the Gibbs energy of a phase. However severely strained crystals, for example, will have superabundant defects and, as a result, will have a significantly increased Gibbs energy. In experiments, aragonite has grown from severely strained calcite in the normal stability field of calcite as a result of this effect. Defects will be ignored in this section.

An important part of crystal chemistry is the consideration of bonding, the cohesive forces between adjacent atoms in the crystal. Unfortunately this approach is not helpful in understanding the activity–composition relations in solid solutions, although, in principle, it is possible to calculate the thermodynamic properties of end-members and solid solutions from the properties of the atoms of the constituent elements and their disposition in the structure. These calculations are impossible for real crystals on the present generation of computers. Thus the approach taken here is simplistic; atoms are considered to be located on sites, points on the lattice, which have fixed geometric relations to the adjacent sites, without specifically considering bonding.

As most of the crystalline solids which will be considered are silicates, the development will be for silicates, although the same logic applies to other groups of minerals which are of interest, for example the spinels and the sulphides. A silicate can be considered as consisting of more or less close-packed oxygens in which some of the interstices are occupied by cations, the nature of the cations depending on the size and shape of the interstices. The proportion of occupied interstices is controlled by charge balance considerations as the formula unit must be charge balanced. For example, forsterite, Mg_2SiO_4 is $2(Mg^{2+})+(Si^{4+})=4(O^{2-})$. In a solid the actual three-dimensional distribution of the oxygens and the different interstices forms a regular lattice because of the 'pull' of the bonding between the oxygens and the cations. Silicon, at least at crustal and upper mantle pressures, occupies an interstice in the middle of four oxygens—silicon is said to be four-fold coordinated to oxygen or simply to be in a tetrahedral site (from the shape of the interstice). Tetrahedral sites are usually large enough to accommodate aluminium and even ferric iron but not the larger cations, titanium, magnesium, ferrous iron and so on.

Another main type of interstice is surrounded by six oxygens; a cation occupying such an interstice is said to be six-fold coordinated to oxygen, or to be in an octahedral site. Such a site can be occupied by aluminium, ferric iron, magnesium, titanium, ferrous iron and, with difficulty, calcium, but is not large enough for sodium or potassium and is too large for silicon. A larger site is required to accommodate sodium and potassium, and calcium often accompanies these larger cations. For octahedral (and larger sites) the description of a site as being n-fold coordinated to oxygens becomes ambiguous because the n oxygens surrounding the site may not all be at the same distance from the cation. For example, calcium prefers an octahedral site which is somewhat open and therefore larger. Such a site may have six oxygens which are nearest the occupying cation, and one or two more oxygens a relatively small distance further away. This is the case for the M2 octahedral site in Ca-rich clinopyroxenes in which calcium is located. An octahedral site can also be effectively smaller if it is distorted. Such distortion will result in the smaller of the usual octahedral cations preferring the site.

In the above, the atoms in the lattice are considered as charged balls located in holes in the close-packed oxygen network. This is obviously an over-simplification, but it does allow a crude consideration of some of the features of sites and site occupancies. There are many factors, apart from 'ionic' size and charge, that control site occupancies. Unfortunately site occupancies cannot be predicted in a quantitative way taking all these factors together—so these other factors need not be considered here.

The fact that several elements can substitute for each other on each site in a mineral means that most silicates form complex solid solutions. Much of the rest of this chapter is concerned with expressing activity–composition relations for end-members in these complex solid solutions.

Mole fractions

In simple solid solutions, for example our phase A involving end-members 1 and 2, we used x_{1A} in:

$$\mu_{1A} = G_{1A} + RT \ln x_{1A}\gamma_{1A} \tag{4.1}$$

where x_{1A} is the proportion of A which consists of 1. For complicated solid solutions, we have to be more careful about how we formulate the mole fraction because the mole fraction in the chemical potential expression is supposed to refer to the way atoms or molecules are mixing in the phase. In a simple solution, say H_2O–CO_2 fluid, there is no problem because it is H_2O and CO_2 molecules which are actually mixing, so the proportion of H_2O in the fluid defines x_{H_2O} in the fluid.

However consider a clinopyroxene which is a binary solid solution between diopside (di) and jadeite (jd), ($CaMgSi_2O_6$–$NaAlSi_2O_6$). Ca and Na are located in the larger M2 site, while Mg and Al are located in the smaller M1 site. A particular composition clinopyroxene can be represented by a block diagram (figure 4.1). Now, if we want to know $x_{CaMgSi_2O_6,cpx}$ we require the way $CaMgSi_2O_6$ and $NaAlSi_2O_6$ are mixing in the solid solution. There are two main ways of considering this:

1: The clinopyroxene consists of a solution of $CaMgSi_2O_6$ and $NaAlSi_2O_6$ *molecules*. Each Ca is considered to be joined to an Mg, and each Na to an Al, and the solid solution is a mixture of these 'dumbells'. In this case $x_{CaMgSi_2O_6,cpx}$ is the proportion of CaMg dumbells, or $x_{di,cpx}$. For the cpx in figure 4.1 this mole fraction is equal to 0·7. This behaviour in a solid solution can be called molecular mixing; it implies complete short-range order. Short-range order reflects the preference of atoms of one element to be adjacent to atoms of another element, in this case the atoms being on different sites.

2:   The clinopyroxene consists of a random mixture of Mg and Al atoms on M1 and a random mixture of Na and Ca atoms on M2. The contribution to the chemical potential of the mixing of cations on each of the sites is the sum of the appropriate mole fraction terms for the sites. Thus, assuming ideal mixing on this basis,

$$\mu_{CaMgSi_2O_6,cpx} = G_{CaMgSi_2O_6,cpx} + RT \ln x_{Ca,M2,cpx} + RT \ln x_{Mg,M1,cpx}$$
$$= G_{CaMgSi_2O_6,cpx} + RT \ln (x_{Ca,M2}x_{Mg,M1})_{cpx} \qquad (4.2)$$

In comparison with (4.1):

$$x_{CaMgSi_2O_6,cpx} = (x_{Ca,M2}x_{Mg,M1})_{cpx}$$

i.e. the product of the mole fractions of the elements in the formula unit on their respective sites. For the cpx in figure 4.1, the mole fraction of

4.1   A block diagram showing the octahedral site occupancies in a $CaMgSi_2O_6$–$NaAlSi_2O_6$ clinopyroxene.

$CaMgSi_2O_6$ on this basis is $(0·7)^2$ or 0·49. This behaviour in a solid solution can be called mixing on sites; it implies an absence of short-range order.

It is also possible to have short-range order on one site, with the Mg atoms on M1 preferring to have Al atoms as nearest neighbours in the clinopyroxene. Complete short-range order of this type would imply the formation of a compound with equal numbers of Al and Mg atoms arranged so that the above preferences are obeyed. For all but 50:50 di–jd solid solutions which have equal numbers of Al and Mg, the clinopyroxene could only be constructed of this ordered compound using domains, small volumes of the compound, seperated by domain boundaries across which the nearest-neighbour preferences are not obeyed.

A simplistic view is adopted that if placing one cation on a site fixes the position of all the other cations on that site, then no mixing term is involved for that site. This would mean that $x_{CaMgSi_2O_6,cpx} = x_{Ca,M2,cpx}$ for complete short-range order of Mg and Al on M1 in the clinopyroxene.

Clearly it is possible for there to be a complete spectrum of behaviour (i.e. amount of short-range order) between these two extremes in different solid solutions, with resulting differences in the formulation of mole fractions. However, a major problem concerning silicates is that there is no simple means of discovering the amount of short-range order in a solid solution, unless it affects the crystallography, by changing the symmetry relations in the crystal. This effect usually appears only when short-range order is almost complete.

For many silicate solid solutions, a choice has to be made whether mole fractions are to be modelled using molecular mixing or mixing on sites. Mixing on sites is the approach usually used unless there are crystallographic or other reasons for using molecular mixing. The usual situations in which strong short-range order is suspected are where cations of quite different size or charge occur in the same sites. For example, if a lattice is accomodating a large cation, the lattice may be less distorted if these large cations do not occur adjacent to each other. If one site contains a larger cation and the other contains a smaller cation, then the lattice may be less distorted if the larger cation is adjacent to the smaller cation. If one site contains a cation which is higher charged, and the other contains a lower charged cation, then it may be energetically more favourable for these two cations to be adjacent to achieve a local charge balance. In all these situations one might expect to get short-range order. However the distortion of the lattice will be small if the larger/smaller or higher/lower charged cations are present in small amounts. In this case short-range order will be minimal and the mixing on sites approach will be appropriate for calculating mole fractions. Thus there should be no problem in using a mixing on sites mole fraction for diopside in di–jd for diopside-rich clinopyroxenes, while the molecular mixing mole fraction may be required for more jadeite-rich compositions. The likely main occurrence of strong short-range order in silicates is in minerals which have tetrahedral sites occupied by Si and Al, like plagioclase feldspars, because of the size and charge difference between these cations.

And so, with a few exceptions, mixing on sites is used in calculating mole fractions of end-members in silicate solid solutions. As the elements in a mineral are distributed among the various sites, we must perform this distribution in order to calculate the mole fractions of these elements on the sites. It is usually possible to distribute most of the elements just on a size criterion. For example, in pyroxenes, Si and sufficient Al are placed on the tetrahedral (T) sites, while the remainder of the Al, plus Fe^{3+} and Ti are placed on the smaller M1 site, while Ca, Na, and Mn are placed on the larger M2 site. However Fe^{2+} (referred to below as Fe for convenience) and Mg are distributed between the M1 and M2 sites. We need to know how to do this for different pyroxene compositions, temperatures and pressures. Most of the appropriate measurements which would allow this to be done have not been made and so approximations are required. One, which will be used for clinopyroxenes and other appropriate phases, is to assume that Fe and Mg are distributed randomly between the two sites, implying:

$$\left(\frac{Fe}{Mg}\right)_{M1} = \left(\frac{Fe}{Mg}\right)_{M2} = \left(\frac{Fe}{Mg}\right)_{px} \tag{4.3}$$

from which the site occupancies can be calculated (see p 76).

A problem can arise with mole fractions formulated by mixing on sites when we require that $x_{1A} = 1$ when phase A consists of pure end-member 1. Consider the mole fraction of calcium Tschermak's molecule (CATS), $CaAlAlSiO_6$:

$$x_{CaAlAlSiO_6,cpx} = (x_{Al,M1} x_{Ca,M2} x_{Al,T} x_{Si,T})_{cpx} \quad \text{(incomplete)}$$

There are two ways of treating CATS depending on the nature of pure CATS:

1: If we assume that the tetrahedral site Al and Si mix randomly on the two sites in the formula unit of pure CATS, then, for pure CATS $x_{Al,M1} = x_{Al,T} = 0.5$, so that the mole fraction of CATS using the above formula is 0.25. To remove this problem, a normalization constant is included in the mole fraction expression to ensure $x_{1A} = 1$ for pure 1. Thus:

$$x_{CaAlAlSiO_6,cpx} = \frac{x_{Al,M1} x_{Ca,M2} x_{Al,T} x_{Si,T}}{(x_{Al,M1} x_{Ca,M2} x_{A1,T} x_{Si,T})_{pure\ CATS}}$$

which for this case is:

$$x_{CaAlAlSiO_6,cpx} = 4 x_{Al,M1} x_{Ca,M2} x_{Al,T} x_{Si,T}$$

2: If on the other hand pure CATS has Si and Al showing strong short-range order on the tetrahedral sites, amounting to Si occupying its own tetrahedral site, T_1, with Al occupying T_2, then:

$$x_{CaAlAlSiO_6,cpx} = x_{Al,M1} x_{Ca,M2} x_{Al,T2} x_{Si,T1}$$

No normalization is required because $x_{Al,T2} = x_{Si,T1} = 1$ in pure ordered CATS. If the mole fraction is required for a pyroxene with a small amount of tetrahedral Al in which there is effectively no short-range order, then the tetrahedral sites T_1 and T_2 are equivalent, and:

$$x_{CaAlAlSiO_6,cpx} = x_{Al,M1} x_{Ca,M2} x_{Al,T} x_{Si,T}$$

No normalization constant is required because the standard state is ordered CATS even though the cpx of interest shows no ordering.

This example emphasizes the need to be clear about the nature of the pure end-member; in this case it makes a difference of a factor of 4 to the mole fraction.

Activity coefficients

The activity coefficients considered here are those that appear in the activity for standard state 1, so $\gamma_{1A} \to 1$ as $x_{1A} \to 1$ (Raoult's law) and $\gamma_{1A} \to$ constant as $x_{1A} \to 0$ (Henry's law). The activity coefficient reflects the dissimilarity of the atoms or molecules being mixed in the solid solution, the more dissimilar they are, the larger the activity coefficient. The activity coefficient can be related to the interactions between the atoms or molecules in the solid solution, usually approximated by the interactions between immediately adjacent (or nearest-neighbour) atoms or molecules. Consider a phase A with atoms of 1 and 2 mixing on a site. An atom of 1 will be surrounded by atoms of 1 and 2 and will interact with them (via bonding, distortion of the lattice, and so on). If this atom of 1 is replaced by an atom of 2, then this will interact in a different way with its neighbours, depending on, for example, the size and charge difference between 1 and 2. The difference between these interactions reflects the dissimilarity between 1 and 2 and controls the magnitude of the activity coefficients. If the sum of these interactions is the same for the atom of 1 as for the atom of 2, then $\gamma_{1A} = 1$ and $\gamma_{2A} = 1$ and the phase is an ideal solution. If an atom of 1 prefers to have neighbours of 2, then we have negative deviations from ideality, $\gamma_{1A} < 1$ and $\gamma_{2A} < 1$, whereas if atoms of 1 prefer to have neighbours of 1 then we have positive deviations from ideality, $\gamma_{1A} > 1$ and $\gamma_{2A} > 1$. In the first case there is a tendency to compound formation, a 12 compound, whereas in the second case there is tendency to unmixing, the formation of a phase consisting predominantly of 1, and a phase consisting predominantly of 2. A good example of compound formation in this sense is the presence of dolomite, $CaMg(CO_3)_2$, between calcite, $CaCO_3$, and magnesite, $MgCO_3$, figure 4.2(a). Unmixing is the usual behaviour in silicate systems, indicating that activity coefficients larger than one are usual in silicate systems. The tendency to unmixing decreases to higher temperature because the lattice, as it becomes more open, tolerates bigger differences between the constituent atoms 1 and 2. This results in the solvus phase diagram (figure 4.2(b)) with a continuous solid solution across the system at high temperature being interrupted by a rapidly expanding solvus to lower temperatures. The top of the solvus cannot be observed in some systems because the solidus, the beginning of melting, is at a lower temperature than the top of the solvus, the broken line in figure 4.2(b).

Phase diagrams like figure 4.2(b) (broken line) can also occur when the two phases have different structures, and here we describe the subsolidus part of the diagram in terms of a miscibility gap, the term solvus being restricted to coexisting phases having the same structure. When a misci-

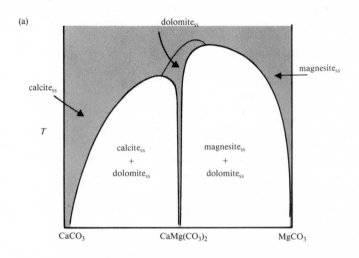

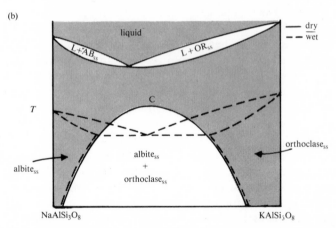

4.2 Schematic phase diagrams showing: (a) an addition compound, dolomite, in the $CaCO_3$–$MgCO_3$ system; and (b) unmixing in the alkali feldspar system. The shading in (b) applies to the phase diagram for dry conditions (full lines). The broken lines define the phase diagram for wet conditions, where the solidus intersects the solvus.

bility gap occurs in a system we cannot make such positive statements about the activity coefficients because the difference between the Gibbs energy of each end-member between the two structures affects the shape of the phase diagram. Usually the narrower the extent of solubility of one phase in another at a particular temperature, the larger is the activity coefficient of the end-member present in a small amount.

Actually putting a numerical value on the activity coefficient of an end-member in a particular phase is a difficult task. Ideally one would be able to look up in tables for the values but very little of the necessary experimental work has yet been done, partly because of experimental difficulties and partly because of the enormous amount of experimental work required to find activity coefficients as functions of composition, temperature, and pressure. We are reduced to making approximations, particularly concerning the form of the composition, temperature, and pressure dependence of the activity coefficients. Once such a form, called a mixing model, has been assumed, the amount of experimental work required to find the constants in the mixing model, and thus the activity coefficients, is much reduced. The problem is that it is not at all clear which of the many mixing models should be used, particularly as it is most unlikely that one mixing model is applicable to all solid solutions, or even to one solid solution over its whole composition range. The simplest approach, apart from assuming ideal mixing, is to use the regular mixing model. The regular model activity coefficients for a binary phase A between atoms or molecules 1 and 2 are:

$$RT \ln \gamma_{1A} = w_{12}(1 - x_{1A})^2$$
$$RT \ln \gamma_{2A} = w_{12}(1 - x_{2A})^2$$

$$(4.4)$$

where w_{12} is the interaction parameter between 1 and 2, and is usually taken to be linear in temperature and pressure, thus $w_{12} = a_{12} + b_{12}T + c_{12}P$, where a, b, and c are constants. w_{12} can be considered as the difference in energy, ε, between 1–1 plus 2–2 nearest neighbours and 1–2 nearest neighbours, or $w_{12} = 2\varepsilon_{12} - \varepsilon_{11} - \varepsilon_{22}$. If 1–1 and 2–2 nearest neighbours are energetically more favourable than 1–2 nearest neighbours, i.e. if they have lower energy, then w_{12} is positive. Unmixing is therefore associated with positive w_{12} values (figure 4.3). Ideal mixing is characterized by $w_{12} = 0$ (i.e. $\gamma_{1A} = 1$). The Henry's law constants are $h_{1A(2)} = h_{2A(1)} = \exp(w_{12}/RT)$.

For more complicated systems, w terms are required for the interactions between each of the atoms or molecules in each of the sites. Thus for the ternary phase A involving end-members 1, 2, and 3:

$$RT \ln \gamma_{1A} = w_{12}x_{2A}(1 - x_{1A}) + w_{13}x_{3A}(1 - x_{1A}) - w_{23}x_{2A}x_{3A}$$
$$RT \ln \gamma_{2A} = w_{12}x_{1A}(1 - x_{2A}) + w_{23}x_{3A}(1 - x_{2A}) - w_{13}x_{1A}x_{3A}$$
$$RT \ln \gamma_{3A} = w_{13}x_{1A}(1 - x_{3A}) + w_{23}x_{2A}(1 - x_{3A}) - w_{12}x_{1A}x_{2A}$$

From these equations we can see that the Henry's law constant for end-member 1 (i.e. when $x_{1A} \to 0$), in a mixture of 2 and 3 is:

$$h_{1A(23)} = \exp[(w_{12}x_{2A} + w_{13}x_{3A} - x_{2A}x_{3A}w_{23})/RT]$$

and so is a function of the proportion of 2 and 3 in A, but not of the proportion of 1 in A. This has particular relevance to using standard state 2; in this case G_{1A}^* would be a function of x_{2A} and x_{3A}.

One of the unfortunate properties of the regular model is that it constrains the thermodynamics of a binary solution to be symmetrical with composition, for example the two Henry's law constants are the same in a binary regular system. This implies that replacing an atom of 1 with 2 in a phase consisting of 1 only takes the same energy as replacing an atom of

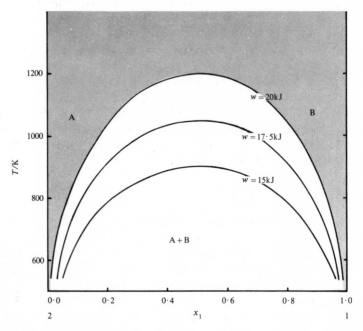

4.3 A T–x diagram showing the decrease in the size of the solvus which corresponds to a decrease in the magnitude of w for the regular solution model. The shading applies only to the $w = 20$ kJ solvus.

2 with 1 in a phase consisting of 2 only, which seems unlikely. There are various ways of removing this symmetry, but these require two (or more) interaction parameters between each pair of atoms, and the number of parameters becomes excessive for complicated solid solutions. For most purposes it will be sufficient to use the regular mixing model when we need to use activity coefficients, although it should be noted that when considering reactions which have a small temperature dependence for their ΔG° that it may be necessary to use more complicated expressions for the activity coefficients. The useful corollary of this is that the

assumption of ideal mixing may not cause appreciable error when considering a reaction which has a large temperature dependence for its ΔG°.

Worked examples 4

4(a)   Calculate the mole fractions of $NaAlSi_2O_6$, $CaMgSi_2O_6$, $Mg_2Si_2O_6$, $Fe_2Si_2O_6$ and $CaAl_2SiO_6$ in the following Ca-rich clinopyroxene:

Oxide	wt %	Oxide	wt %
SiO_2	54·09	MnO	0·09
TiO_2	0·28	MgO	16·96
Al_2O_3	1·57	CaO	21·10
Fe_2O_3	0·74	Na_2O	1·37
Cr_2O_3	2·03	K_2O	0·15
FeO	1·47		

The first step is to recalculate the analysis.

	wt %	wt%/MW (a)	(a)$\times N_O$	(a)$\times N_M$ (b)	(b)$\times F$ (c)	$2N_O$(c)/N_M
SiO_2	54·09	0·9001	1·8002	0·9001	1·9641	7·8564
TiO_2	0·28	0·0035	0·0070	0·0035	0·0076	0·0304
Al_2O_3	1·57	0·0154	0·0462	0·0308	0·0672	0·2016
Fe_2O_3	0·74	0·0046	0·0138	0·0092	0·0201	0·0603
Cr_2O_3	2·03	0·0134	0·0402	0·0268	0·0585	0·1755
FeO	1·47	0·0205	0·0205	0·0205	0·0447	0·0894
MnO	0·09	0·0013	0·0013	0·0013	0·0028	0·0056
MgO	16·96	0·4206	0·4206	0·4206	0·9178	1·8356
CaO	21·10	0·3762	0·3762	0·3762	0·8209	1·6418
Na_2O	1·37	0·0221	0·0221	0·0442	0·0964	0·0964
K_2O	0·15	0·0016	0·0016	0·0032	0·0070	0·0070
			2·7497			12·0000

$F = 6/2 \cdot 7497 = 2 \cdot 1821$

In the above scheme, MW stands for molecular weight of the appropriate oxide, N_M for the number of metal atoms in the oxide formula unit, N_O for the number of oxygens in the oxide formula unit. The column labelled (c) gives the cation per six oxygens values. The last column is a check of the arithmetic; the sum of the cation values times $2N_O/N_M$ should equal twice the number of oxygens in the formula unit of the mineral (for example, twelve for pyroxenes). Although four decimal places are retained throughout the recalculation, it does not imply this precision to the analysis. Three significant figures would be a generous estimate of precision on the results. That number of decimal places are retained to avoid rounding errors in the calculations. The analysis can be written as:

$$(K_{0.007}Na_{0.096}Ca_{0.821}Mg_{0.918}Mn_{0.003}Fe^{2+}_{0.045}Cr_{0.059}Fe^{3+}_{0.020}Ti_{0.008}Al_{0.031})$$
$$(Al_{0.036}Si_{1.964})O_6$$

with Al being used to fill the tetrahedral site. As the two tetrahedral (T) sites are equivalent in the formula unit, then:

$$x_{Al,T} = 0.0362/2 = 0.018 \quad x_{Si,T} = 1.964/2 = 0.982$$

Placing the trivalent cations plus Ti on the smaller M1 site:

$$x_{Al,M1} = 0.031 \quad x_{Fe3,M1} = 0.020 \quad x_{Cr,M1} = 0.059 \quad x_{Ti,M1} = 0.008$$

Placing the monovalent cations plus Ca and Mn on the larger M2 site:

$$x_{Ca,M2} = 0.821 \quad x_{Na,M2} = 0.096 \quad x_{K,M2} = 0.007 \quad x_{Mn,M2} = 0.003$$

The distribution of Fe^{2+} (shortened to just Fe below) and Mg between M1 and M2 is achieved using the assumption that:

$$\frac{x_{Fe,M1}}{x_{Mg,M1}} \Big/ \frac{x_{Fe,M2}}{x_{Mg,M2}} = a \quad \text{where } a = 1 \text{ usually}$$

This leads to:

$$\left(\frac{x_{Fe}}{x_{Mg}}\right)_{M1} = \left(\frac{x_{Fe}}{x_{Mg}}\right)_{M2} = \left(\frac{x_{Fe}}{x_{Mg}}\right)_{cpx} = z$$

Now:

$$x_{Fe,M1} = 1 - x_{Mg,M1} - x_{Fe3,M1} - x_{Al,M1} - x_{Cr,M1} - x_{Ti,M1}$$
$$= y - x_{Mg,M1}$$

where y is used to represent one minus the sum of the trivalent and quadrivalent site occupancies. Substituting back:

$$\frac{y - x_{Mg,M1}}{x_{Mg,M1}} = z$$

Re-arranging:

$$x_{Mg,M1} = \frac{y}{z+1} \quad \text{and} \quad x_{Fe,M1} = \frac{zy}{z+1}$$

also

$$x_{Mg,M2} = c_{Mg} - \frac{y}{z+1} \quad \text{and} \quad x_{Fe,M2} = z\left(c_{Mg} - \frac{y}{z+1}\right)$$

where c_{Mg} is the amount of recalculated Mg in the analysis. In this case:

$$z = 0.0487 \quad y = 0.8825 \quad c_{Mg} = 0.9178$$

Therefore:

$$x_{Mg,M1} = 0.842 \qquad x_{Fe,M1} = 0.041$$
$$x_{Mg,M2} = 0.076 \qquad x_{Fe,M2} = 0.0037$$

The arithmetic can now be checked by making sure that the starting assumption is obtained:

$$\frac{x_{Fe,M1}}{x_{Mg,M1}} \bigg/ \frac{x_{Fe,M2}}{x_{Mg,M2}} = \frac{0.041}{0.842} \bigg/ \frac{0.0037}{0.076} = 1$$

Mixing on sites should be appropriate for the calculation of the mole fractions for this clinopyroxene. Therefore:

$$x_{CaMgSi_2O_6,cpx} = (0.821)(0.842)(0.982)^2 = 0.667$$
$$x_{Mg_2Si_2O_6,cpx} = (0.842)(0.076)(0.982)^2 = 0.062$$
$$x_{Fe_2Si_2O_6,cpx} = (0.041)(0.0037)(0.982)^2 = 0.00015$$
$$x_{CaAl_2SiO_6,cpx} = 4(0.821)(0.031)(0.018)(0.982) = 0.0018$$

It is worthwhile noting that the effect of changing the Fe^{2+}–Mg site distribution assumption (i.e. making $a \neq 1$) is to change the calculated values for $x_{Fe,M1}$ and $x_{Fe,M2}$ much more than the values of $x_{Mg,M1}$ and $x_{Mg,M2}$ for this Mg-rich clinopyroxene. This means that uncertainties in the mole fractions of iron end-members are much larger than for the mole fractions of the magnesium end-members. Similarly the normal analytical uncertainties on the SiO_2 wt% value seriously affects $x_{Al,T}$ as this is obtained by difference from $x_{Si,T}$. Thus $x_{Al,M1}$ is also very uncertain because it is obtained by difference from c_{Al} and $x_{Al,T}$. Therefore the mole fractions of aluminium-containing end-members are uncertain. An attempt at putting values on these uncertainties will be made later.

4(b) Calculate the mole fractions of $Mg_7Si_8O_{22}(OH)_2$ (cummingtonite), $Ca_2Mg_5Si_8O_{22}(OH)_2$ (tremolite), $Ca_2Mg_3Al_2Si_6Al_2O_{22}(OH)_2$ (tschermakite), $NaCa_2Mg_5Si_7AlO_{22}(OH)_2$ (edenite) and $NaCa_2Mg_4Al Si_6Al_2O_{22}(OH)_2$ (pargasite) in the following hornblende:

Oxide	wt%	Oxide	wt%
SiO_2	42·05	FeO	6·34
TiO_2	1·48	MgO	14·91
Al_2O_3	14·69	CaO	12·83
Fe_2O_3	3·21	Na_2O	2·41

The analysis is recalculated on 23 oxygens, assuming 1 H_2O/formula unit.

Oxide	wt%	wt%/MW (a)	$(a) \times N_O$	$(a) \times N_M$ (b)	$(b) \times F$ (c)	$2N_O(c)/N_M$
SiO_2	42·05	0·6998	1·3996	0·6998	6·0625	24·2499
TiO_2	1·48	0·0185	0·0370	0·0185	0·1603	0·6412
Al_2O_3	14·69	0·1441	0·4323	0·2882	2·4967	7·4902
Fe_2O_3	3·21	0·0201	0·0603	0·0402	0·3483	1·0449
FeO	6·34	0·0882	0·0882	0·0882	0·7641	1·5282
MgO	14·91	0·3698	0·3698	0·3698	3·2037	6·4074
CaO	12·83	0·2288	0·2288	0·2288	1·9821	3·9642
Na_2O	2·41	0·0389	0·0389	0·0778	0·6740	0·6740
			2·6549			46·0000

$F = 23/2·6549 = 8·6632$

In hornblendes there are two types of tetrahedral sites, T_1 and T_2 in proportion 4:4, with Al only allowed on the first of these. The required Al to fill T_1 is $8 - 6·0625 = 1·9375$. Therefore:

$$x_{Si,T2} = 1 \quad x_{Si,T1} = 2·0625/4 = 0·516$$
$$x_{Al,M1} = 1·9375/4 = 0·484$$

There are four octahedral sites, M1, M2, M3 and M4, in proportion 2:2:1:2. Trivalent cations prefer the smaller M2 site, while Ca and Na prefer the larger M4 site. Therefore:

$$x_{Al,M2} = (2·4967 - 1·9375)/2 = 0·280$$
$$x_{Fe3,M2} = 0·3483/2 = 0·174$$
$$x_{Ti,M2} = 0·1603/2 = 0·080$$

and:

$$x_{Ca,M4} = 1·9821/2 = 0·991$$
$$2x_{Na,M4} = 7 - (0·7641 + 3·2037 + 0·5592 + 0·3483 + 0·1603 + 1·9821)$$
$$= -0·018$$

In other words, there is no room for Na on the M4 site. Therefore:

$$x_{Na,A} = 0·674$$
$$x_{o,A} = 0·326$$

The last being the mole fraction of vacancies on the A site.

The same method of distributing Fe and Mg between M1–4 is used for

amphiboles as for pyroxenes (WE 4(a)). Treating M1 and M3 as indistinguishable (M13), then:

$$\left(\frac{x_{Fe}}{x_{Mg}}\right)_{M13} = \left(\frac{x_{Fe}}{x_{Mg}}\right)_{M2} = \left(\frac{x_{Fe}}{x_{Mg}}\right)_{M4} = \left(\frac{x_{Fe}}{x_{Mg}}\right)_{amph} = r$$

Now:

$$x_{Fe,M13} = 1 - x_{Mg,M13}$$

$$x_{Fe,M2} = s - x_{Mg,M2} \quad \text{where} \quad s = 1 - x_{Fe3,M2} - x_{Al,M2} - x_{Ti,M2}$$

$$x_{Fe,M4} = t - x_{Mg,M4} \quad \text{where} \quad t = 1 - x_{Ca,M4}$$

As before:

$$x_{Mg,M13} = \frac{1}{r+1} \quad x_{Fe,M13} = 1 - x_{Mg,M13} = \frac{r}{r+1}$$

$$x_{Mg,M2} = \frac{s}{r+1} \quad x_{Fe,M2} = s - x_{Mg,M2} = \frac{rs}{r+1}$$

$$x_{Mg,M4} = \frac{t}{r+1} \quad x_{Fe,M4} = t - x_{Mg,M4} = \frac{rt}{r+1}$$

Substituting:

$$r = 0 \cdot 2385, \quad s = 0 \cdot 466 \quad \text{and} \quad t = 0 \cdot 0090$$

and therefore:

$$x_{Mg,M13} = 0 \cdot 807 \quad x_{Fe,M13} = 0 \cdot 193$$

$$x_{Mg,M2} = 0 \cdot 376 \quad x_{Fe,M2} = 0 \cdot 090$$

$$x_{Mg,M4} = 0 \cdot 0073 \quad x_{Fe,M4} = 0 \cdot 0017$$

Using mixing on sites for calculating the mole fractions:

(a) $x_{Mg_7Si_8O_{22}(OH)_2} = (0 \cdot 326)(0 \cdot 807)^3(0 \cdot 376)^2(0 \cdot 0073)^2(1)^4(0 \cdot 516)^4$

$$= 9 \cdot 151 \times 10^{-8}$$

(b) $x_{Ca_2Mg_5Si_8O_{22}(OH)_2} = (0 \cdot 326)(0 \cdot 807)^3(0 \cdot 376)^2(0 \cdot 991)^2(1)^4(0 \cdot 516)^4$

$$= 0 \cdot 00169$$

(c) $x_{Ca_2Mg_3Al_2Si_6Al_2O_{22}(OH)_2} = 16(0 \cdot 326)$

$$\times (0 \cdot 807)^3(0 \cdot 280)^2(0 \cdot 991)^2(1)^4(0 \cdot 516)^2(0 \cdot 484)^2$$

$$= 0 \cdot 0132$$

(d) $x_{NaCa_2Mg_5Si_7AlO_{22}(OH)_2} = 9 \cdot 48(0 \cdot 674)$

$$\times (0 \cdot 807)^3(0 \cdot 376)^2(0 \cdot 991)^2(1)^4(0 \cdot 516)^3(0 \cdot 484)$$

$$= 0 \cdot 0310$$

(e) $x_{NaCa_2Mg_4AlSi_6Al_2O_{22}(OH)_2} = 64(0\cdot674)(0\cdot807)^3(0\cdot376)(0\cdot280)$

$$\times (1)^4(0\cdot516)^2(0\cdot484)^2$$

$$= 0\cdot149$$

Note that in (a)–(c), $x_{o,A}$ must be included in the mole fraction, and that in (c)–(e), normalization constants are required. If the analysis had contained fluorine, then each mole fraction would have to include a x_{OH}^2 term to account for the mixing of OH and F on the hydroxyl site. Note that we could not calculate the mole fraction of any alkali amphibole end-member which has M4 site Na (for example, glaucophane, Na_2Mg_3-$Al_2Si_8O_{22}(OH)_2$) because the amphibole of interest contains no M4 site Na.

In these calculations we have assumed that the mixing on sites formulation applies to each of the sites. It is possible that, for example, the mixing on sites formulation is inapplicable for the mixing of the Al and Si on T1 due to substantial short-range order effects; in which case the tetrahedral site terms (and the appropriate normalization constants) should not be included in the mole fractions.

4(c)   Calculate the mole fractions of $KAl_3Si_3O_{10}(OH)_2$ (muscovite), Na-$Al_3Si_3O_{10}(OH)_2$ (paragonite), $KMg_3AlSi_3O_{10}(OH)_2$ (phlogopite), $KFe_3AlSi_3O_{10}(OH)_2$ (annite) and $KMg_3AlSi_3O_{10}(F)_2$ (fluorphlogopite) in the biotite (BI):

$$Na_{0\cdot03}K_{0\cdot97}Mg_{1\cdot65}Fe_{0\cdot80}Al_{1\cdot50}Si_{2\cdot90}O_{10}OH_{1\cdot7}F_{0\cdot3}$$

and in the muscovite (MU):

$$Na_{0\cdot08}K_{0\cdot92}Mg_{0\cdot23}Fe_{0\cdot07}Al_{2\cdot80}Si_{3\cdot00}O_{10}OH_{1\cdot9}F_{0\cdot1}$$

The distribution of cations in muscovites and biotites can be considered in the same way. They involve an alkali site (A) on which the Na and K are located. There are two octahedral sites, M1 and M2, in proportion $1:2$. In muscovite, the octahedral Al is located on M2 with much of the M1 site vacant. We have to assume that the octahedral Al and vacancies are distributed in the same way in biotite. There are four tetrahedral sites (T) across which the Si and tetrahedral Al are evenly distributed.

Thus for the BI

$$x_{Na,A} = 0\cdot03 \qquad\qquad x_{K,A} = 0\cdot97$$

$$x_{OH,V} = 1\cdot7/2 = 0\cdot85 \qquad x_{F,V} = 0\cdot3/2 = 0\cdot15$$

$$x_{Si,T} = 2\cdot90/4 = 0\cdot725 \qquad x_{Al,T} = 1 - 0\cdot725 = 0\cdot275$$

$$x_{Al,M2} = [1\cdot5 - 4(0\cdot275)]/2 = 0\cdot2$$

$$x_{o,M1} = 3 - [1\cdot65 - 0\cdot80 - 2(0\cdot2)] = 0\cdot15$$

Assuming that the Mg and Fe are distributed across M1 and M2 according to:

$$\left(\frac{x_{Fe}}{x_{Mg}}\right)_{M1} = \left(\frac{x_{Fe}}{x_{Mg}}\right)_{M2} = \left(\frac{x_{Fe}}{x_{Mg}}\right)_{BI} = r$$

Now:

$$x_{Fe,M1} = s - x_{Mg,M1} \quad \text{with} \quad s = 1 - x_{o,M1}$$

$$x_{Fe,M2} = t - x_{Mg,M2} \quad \text{with} \quad t = 1 - x_{Al,M2}$$

So:

$$x_{Mg,M1} = \frac{s}{1+r} \qquad x_{Fe,M1} = \frac{rs}{1+r}$$

$$x_{Mg,M2} = \frac{t}{1+r} \qquad x_{Fe,M2} = \frac{rt}{1+r}$$

Substituting into these:

$$r = 0 \cdot 485, \quad s = 0 \cdot 85 \quad \text{and} \quad t = 0 \cdot 80$$

and:

$$x_{Mg,M1} = 0 \cdot 572 \qquad x_{Fe,M1} = 0 \cdot 278$$

$$x_{Mg,M2} = 0 \cdot 539 \qquad x_{Fe,M2} = 0 \cdot 261$$

The mole fractions on a mixing on sites basis are:

$$x_{KAl_3Si_3O_{10}(OH)_2,BI} = 9 \cdot 48(0 \cdot 97)(0 \cdot 15)(0 \cdot 2)^2(0 \cdot 275)(0 \cdot 725)^3(0 \cdot 85)^2$$
$$= 0 \cdot 00418$$

$$x_{NaAl_3Si_3O_{10}(OH)_2,BI} = 9 \cdot 48(0 \cdot 03)(0 \cdot 15)(0 \cdot 2)^2(0 \cdot 275)(0 \cdot 725)^3(0 \cdot 85)^2$$
$$= 0 \cdot 000129$$

$$x_{KMg_3AlSi_3O_{10}(OH)_2,BI} = 9 \cdot 48(0 \cdot 97)(0 \cdot 572)(0 \cdot 539)^2(0 \cdot 275)(0 \cdot 725)^3(0 \cdot 85)^2$$
$$= 0 \cdot 116$$

$$x_{KFe_3AlSi_3O_{10}(OH)_2,BI} = 9 \cdot 48(0 \cdot 97)(0 \cdot 278)(0 \cdot 261)^3(0 \cdot 275)(0 \cdot 725)^3(0 \cdot 85)^2$$
$$= 0 \cdot 0132$$

$$x_{KMg_3AlSi_3O_{10}(F)_2,BI} = 9 \cdot 48(0 \cdot 97)(0 \cdot 572)(0 \cdot 539)^2(0 \cdot 275)(0 \cdot 725)^3(0 \cdot 15)^2$$
$$= 0 \cdot 00360$$

In each case the normalization constant, 9·48, arises from mixing of Si and Al in the T site in the pure end-member. In the same way for the

MU:

$$x_{Na,A} = 0.08 \qquad x_{K,A} = 0.92$$

$$x_{OH,V} = 0.95 \qquad x_{F,V} = 0.05$$

$$x_{Si,T} = 0.75 \qquad x_{Al,T} = 0.25$$

$$x_{Al,M2} = 0.90$$

$$x_{o,M1} = 0.90$$

$$x_{Mg,M1} = 0.0767 \qquad x_{Fe,M1} = 0.0233$$

$$x_{Mg,M2} = 0.0767 \qquad x_{Fe,M2} = 0.0233$$

The mole fractions on a mixing on sites basis are:

$$x_{KAl_3Si_3O_{10}(OH)_2,MU} = 0.605$$

$$x_{NaAl_3Si_3O_{10}(OH)_2,MU} = 0.0526$$

$$x_{KMg_3AlSi_3O_{10}(OH)_2,MU} = 0.000375$$

$$x_{KFe_3AlSi_3O_{10}(OH)_2,MU} = 1.050 \times 10^{-5}$$

$$x_{KMg_3AlSi_3O_{10}(F)_2,MU} = 1.038 \times 10^{-6}$$

4(d) Calculate the mole fractions for:

(1) $Mg_3Si_4O_{10}(OH)_2$ (talc) in the talc (TA): $Mg_{2.7}Fe_{0.2}Al_{0.2}Si_{3.9}O_{10}(OH)_2$.

(2) $CaAl_2Si_2O_8$ (anorthite) and $NaAlSi_3O_8$ (albite) in the plagioclase (PL): $Ca_{0.4}Na_{0.6}Al_{1.4}Si_{2.6}O_8$.

(3) $Mg_3Al_2Si_3O_{12}$ (pyrope), $Ca_3Al_2Si_3O_{12}$ (grossularite) and $Ca_3Cr_2Si_3O_{12}$ (uvarovite) in the garnet (G): $Ca_{1.0}Fe_{1.8}Mg_{0.2}Al_{1.6}Cr_{0.4}Si_3O_{12}$.

(4) Mg_2SiO_4 (forsterite), Fe_2SiO_4 (fayalite) and $CaMgSiO_4$ (monticellite) in the olivine (OL): $Mg_{1.6}Fe_{0.3}Ca_{0.1}SiO_4$.

(5) Fe_3O_4 (magnetite) and Fe_2TiO_4 (ulvospinel) in the spinel (SP): $Fe_{2.4}Ti_{0.6}O_4$.

(6) $Ca_2Al_3Si_3O_{12}(OH)$ (clinozoisite) and $Ca_2Fe3Al_2Si_3O_{12}(OH)$ (epidote) in the epidote (EP): $Ca_2Al_{2.6}Fe3_{0.4}Si_3O_{12}(OH)$.

(7) $Mg_5Al_2Si_3O_{10}(OH)_8$ (clinochlore) and $Mg_4Al_4Si_2O_{10}(OH)_8$ (corundophyllite) in the chlorite (CHL): $Mg_{2.7}Fe_{1.9}Al_{2.8}Si_{2.6}O_{10}(OH)_8$.

(Fe stands for Fe^{2+} and Fe3 for Fe^{3+} in the above.)

(1) Talc can be considered to have four identical tetrahedral sites and three identical octahedral sites. Therefore, for mixing on sites:

$$x_{Mg_3Si_4O_{10}(OH)_2,TA} = \left(\frac{2.7}{3}\right)^3 \left(\frac{3.9}{4}\right)^4 = 0.659$$

(2) Experimental evidence suggests that short-range order is important

in plagioclase feldspars. Thus molecular mixing should be used in calculating the mole fractions. Thus:

$$x_{NaAlSi_3O_8,PL} = 0\cdot6 \quad \text{and} \quad x_{CaAl_2Si_2O_8,PL} = 0\cdot4$$

(3)   Garnets have three identical tetrahedral sites and three eight-fold coordinated sites. Therefore, for mixing on sites:

$$x_{Mg_3Al_2Si_3O_{12},G} = \left(\frac{0\cdot2}{3}\right)^3 \left(\frac{1\cdot6}{2}\right)^2 (1)^3 = 0\cdot000190$$

$$x_{Ca_3Al_2Si_3O_{12},G} = \left(\frac{1\cdot0}{3}\right)^3 \left(\frac{1\cdot6}{2}\right)^2 (1)^3 = 0\cdot0237$$

$$x_{Ca_3Cr_2Si_3O_{12},G} = \left(\frac{1\cdot0}{3}\right)^3 \left(\frac{0\cdot4}{2}\right)^2 (1)^3 = 0\cdot00148$$

(4)   Olivine has one tetrahedral site and two octahedral sites, M1 and M2. The Ca is located on the larger M2 site. Thus:

$$x_{Ca,M2} = 0\cdot1$$

Assuming that the Fe and Mg are distributed between M1 and M2 according to:

$$\left(\frac{x_{Fe}}{x_{Mg}}\right)_{M1} = \left(\frac{x_{Fe}}{x_{Mg}}\right)_{M2} = \left(\frac{x_{Fe}}{x_{Mg}}\right)_{OL}$$

then, in the same way as in the previous worked examples:

$$x_{Mg,M1} = 0\cdot842 \qquad x_{Fe,M1} = 0\cdot158$$
$$x_{Mg,M2} = 0\cdot758 \qquad x_{Fe,M1} = 0\cdot142$$

Therefore, by mixing on sites:

$$x_{Mg_2SiO_4,OL} = (0\cdot842)(0\cdot758) = 0\cdot638$$
$$x_{Fe_2SiO_4,OL} = (0\cdot158)(0\cdot142) = 0\cdot0224$$
$$x_{CaMgSiO_4,OL} = (0\cdot1)(0\cdot842) = 0\cdot0842$$

Note that the mole fraction of larnite Ca_2SiO_4 is undefined because Ca occurs only on M2.

(5)   Some evidence suggests that short-range order is very important in spinels. In this case, for molecular mixing:

$$x_{Fe_3O_4,SP} = 0\cdot4 \quad \text{and} \quad x_{Fe_2TiO_4,SP} = 0\cdot6$$

(6)   There are three identical tetrahedral sites and two Ca-bearing sites in epidote. Of the three Al-bearing sites only one readily accepts Fe3. Therefore, for mixing on sites:

$$x_{Ca_2Al_3Si_3O_{12}(OH),EP} = (1)^2(0\cdot6)(1)^3 = 0\cdot6$$
$$x_{Ca_2Al_2Fe3Si_3O_{12}(OH),EP} = (1)^2(0\cdot4)(1)^3 = 0\cdot4$$

(7)   Chlorite can be considered as having four identical tetrahedral sites and six identical octahedral sites. Therefore, for mixing on sites:

$$x_{Mg_5Al_2Si_3O_{10}(OH)_8,CHL} = \frac{\left(\frac{2\cdot7}{6}\right)^5\left(\frac{1\cdot4}{6}\right)\left(\frac{1\cdot4}{4}\right)\left(\frac{2\cdot6}{4}\right)^3}{\left(\frac{5}{6}\right)^5\left(\frac{1}{6}\right)\left(\frac{1}{4}\right)\left(\frac{3}{4}\right)^3} = 0\cdot0586$$

$$x_{Mg_4Al_4Si_2O_{10}(OH)_8,CHL} = \frac{\left(\frac{2\cdot7}{6}\right)^4\left(\frac{1\cdot4}{6}\right)^2\left(\frac{1\cdot4}{4}\right)^2\left(\frac{2\cdot6}{4}\right)^2}{\left(\frac{4}{6}\right)^4\left(\frac{2}{6}\right)^2\left(\frac{2}{4}\right)^2\left(\frac{2}{4}\right)^2} = 0\cdot0842$$

The denominator in each case is the normalization constant.

4(e)   Calculate expressions for the activity coefficients of the end-members in a binary system from the solvus shown in figure 4.4(a).

As the solvus is symmetrical, one way of calculating the activity coefficients is to use the regular mixing model. Note that the calculated activity coefficients are most likely to be different if another mixing model is used. In other words, the activity coefficients will be model-dependent.

Below the top of the solvus, the phases are called A and B even though they have the same structure (e.g. belong to the same mineral group). The equilibrium relations for the coexistence of A and B are:

$$\mu_{1A} = \mu_{1B} \quad \text{and} \quad \mu_{2A} = \mu_{2B}$$

Only one of these is required because the solvus is symmetric. Expanding the chemical potentials using standard state 1 we have:

$$0 = \mu_{1A}^\circ - \mu_{1B}^\circ + RT\ln\left(\frac{x_{1A}}{x_{1B}}\right) + RT\ln\gamma_{1A} - RT\ln\gamma_{1B}$$

As A and B have the same structure then $\mu_{1A}^\circ = \mu_{1B}^\circ$ so $\mu_{1A}^\circ - \mu_{1B}^\circ = 0$. Substituting the regular model equations for the activity coefficients:

$$0 = RT\ln\left(\frac{x_{1A}}{x_{1B}}\right) + w(1-x_{1A})^2 - w(1-x_{1B})^2$$

As the solvus is symmetrical $x_{1A} = 1 - x_{1B}$, so substituting $x = x_{1A} = 1 - x_{1B}$ and re-arranging:

$$0 = RT\ln\left(\frac{x}{1-x}\right) + w(1-2x)$$

Re-arranging:

$$w = \frac{RT\ln\left(\frac{x}{1-x}\right)}{2x-1}$$

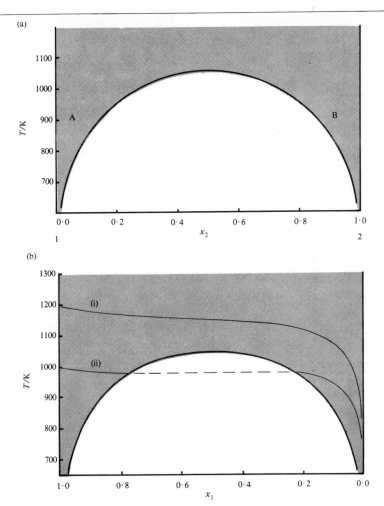

4.4 The $T-x$ diagram, (a), shows a solvus from which activity coefficients are calculated in WE 4(e) assuming that the regular solution model is applicable. The $T-x$ diagram, (b), shows the interaction of two reactions (i) and (ii), involving end-member 1 with this solvus.

Substituting the solvus composition at each temperature:

T/K	x	w/kJ
1000	0·270	18
900	0·135	19
800	0·070	20
700	0·033	21

From this information w appears to be linear in temperature. Applying the method of putting a straight line through two points (p. 263):

$$\frac{w-18}{21-18} = \frac{T-1000}{700-1000}$$

Re-arranging:

$$w = 28 - 0 \cdot 01\,T\,\text{kJ}$$

Thus the activity coefficients of the end-members in this system, *assuming* the regular solution model for the activity coefficients, are:

$$RT \ln \gamma_{1A} = (28 - 0 \cdot 01\,T)(1 - x_{1A})^2$$
$$RT \ln \gamma_{2A} = (28 - 0 \cdot 01\,T)(1 - x_{2A})^2$$

and the same equations for phase B. If the position of the solvus is a function of pressure, then a pressure dependence is required in the w term. If the solvus of interest is asymmetric, then a two-parameter mixing model like the van Laar model is required. The two equilibrium relations are now required to calculate the two mixing parameters at each temperature from the compositions of the limbs of the solvus. The calculation is complicated because the two equations have to be solved iteratively.

4(f)   Use the activity coefficients calculated in the last worked example to calculate the position of a reaction involving one molecule of end-member 1 in binary phase A on the right-hand side (high temperature side) of the reaction as a function of the composition of the phase if ΔG° for the reaction is (a) $60 - 0 \cdot 05\,T$ kJ and (b) $50 - 0 \cdot 05\,T$ kJ and assuming unit activities for the other phases in the reaction.

The equilibrium relation for part (a) is:

$$0 = 60 - 0 \cdot 05\,T + RT \ln x_{1A} + (28 - 0 \cdot 01\,T)(1 - x_{1A})^2$$

and for part (b) the 60 is replaced by 50. Re-arranging this equation:

$$T = \frac{60 + 28(1 - x_{1A})^2}{0 \cdot 05 - R \ln x_{1A} + 0 \cdot 01(1 - x_{1A})^2}$$

The analogous equation for part (b) involves replacing 60 by 50 in this equation. Substitution of a series of x_{1A} values into the two equations gives the two T–x lines, (i) and (ii) in figure 4.4(b). These lines are analogous to the line in figure 3.8 which was calculated assuming ideal mixing. The main effect of positive activity coefficients for end-members on the high temperature side of the reaction is to flatten the reaction line. The larger the activity coefficients the more pronounced the effect. The line is horizontal across the solvus if the solvus is intersected. The experimentally determined positions of dehydration and decarbonation reactions are flattened on $T-x_{CO_2}$ diagrams like (i) in figure 4.4(b) rather than like the line in figure 3.8. This indicates that H_2O-CO_2 fluids are positively non-ideal. In fact there is a solvus between H_2O and CO_2 at low temperatures, with its top at 550 K at 2 kbar.

Problems 4

4(a)   Calculate the mole fractions of $Mg_2Si_2O_6$ (enstatite), $Fe_2Si_2O_6$ (ferrosilite), $CaCrAlSiO_6$ (Cr–CATS), $Mg_{1.5}AlSi_{1.5}O_6$ ('pyrope') in the following orthopyroxene:

Oxide	wt%
SiO_2	57·73
Al_2O_3	0·95
Fe_2O_3	0·42
Cr_2O_3	0·46
FeO	3·87
MgO	36·73
CaO	0·23

4(b)   Calculate the mole fractions of $Ca_2Mg_5Si_8O_{22}(OH)_2$ (tremolite), $Na_2CaMg_5Si_8O_{22}(OH)_2$ (richterite), $Na_2Mg_3Al_2Si_8O_{22}(OH)_2$ (glaucophane), $Mg_7Si_8O_{22}(OH)_2$ (cummingtonite), $NaCa_2Mg_4Al-Si_6Al_2O_{22}(OH)_2$ (pargasite) and $Na_2Mg_3Al_2Si_8O_{22}(F)_2$ (fluor-glaucophane) in the alkali amphibole: $Na_{1.9}Ca_{0.2}Mg_{2.6}Fe_{0.5}Al_{2.1}-Si_{7.8}O_{22}(OH_{1.7}F_{0.3})$.

4(c)   Calculate the mole fractions of
   (1) Mg_2SiO_4 (forsterite) and Fe_2SiO_4 (fayalite) in the olivine (OL): $Mg_{0.8}Fe_{1.2}SiO_4$.
   (2) $Mn_3Al_2Si_3O_{12}$ (spessartine), $Fe_3Al_2Si_3O_{12}$ (almandine), $Ca_3Fe_2-Si_3O_{12}$ (andradite) in the garnet (G): $Mn_{0.2}Fe_{0.3}Mg_{0.2}Ca_{2.3}Al_{0.9}-Fe3_{1.1}Si_3O_{12}$.
   (3) $NaAlSi_3O_8$ (albite), $CaAl_2Si_2O_8$ (anorthite) and $KAlSi_3O_8$ (orthoclase) in the plagioclase (PL): $K_{0.2}Na_{0.7}Ca_{0.1}Al_{1.1}Si_{2.9}O_8$.
   (4) $Fe_5Al_2Si_3O_{10}(OH)_8$ (ferroclinochlore) and $Fe_4Al_4Si_2O_{10}(OH)_8$ (ferrocorundophyllite) in the chlorite: $Mg_{2.3}Fe_{2.5}Al_{2.4}Si_{2.8}-O_{10}(OH)_8$.

References

Wood, B. J. and Fraser, D. G., 1976, *Elementary Thermodynamics for Geologists;* Oxford University Press, London: Chapter 3 contains a more mathematical treatment of activity–composition relations in solids.

McKie, D. and McKie, C., 1973, *Crystalline Solids,* Nelson, London: Chapter 10 is further reading on crystal chemistry (bonding etc.).

Wood, B. J., 1977, Experimental determination of the mixing properties of solid solutions; *and*

Newton, R. C., 1977, Thermochemistry of garnets and aluminous pyroxenes in the CMAS system. In D. G. Fraser (Ed.) *Thermodynamics in Geology,* Reidel, Dordrecht: Advanced reading on the thermodynamics of solids.

Ulbrich, H. H. and Waldbaum, D. R., 1976, Structural and other contributions to the third-law entropies of silicates. *Geochem. Cosmochim. Acta,* **40,** 1–24: Source of information on site distributions in silicates.

Chapter 5

Thermodynamics of Fluids

Fluid is a term which we will use to cover gases and liquids. Many assemblages in rocks formed in the presence of a fluid. In the case of metamorphic rocks and most sedimentary rocks, this fluid has long since disappeared from the rock and the only evidence we have of its previous existence is in the mineralogy or in fluid inclusions in the minerals. In the case of igneous rocks, the fluid involved was the magma, a silicate liquid. For lavas, this fluid may be still present in the form of quenched liquid (glass) although its composition will have been changed by any crystallization of minerals, the amount of the change depending on the amount of crystallization. For plutonic igneous rocks, the composition of the liquid which crystallized the assemblages can only be inferred, particularly if cumulus processes have operated. In all the above situations, the composition and properties of the 'missing' fluid are very important to understanding the processes that formed the rock as we now observe it. We can discover much about the fluid from the compositions of the minerals in the assemblage, with the usual assumption of frozen-in equilibrium, using the equilibrium relation for reactions involving end-members of the fluid, if we know the activity–composition relations in that fluid.

The fluids of interest here can be divided into aqueous fluids and silicate liquids. Silicate liquids will be considered later. The aqueous fluids of interest are natural waters at or near the Earth's surface today, fluids still present in unlithified sediments, and the fluids originally present in sedimentary and metamorphic rocks, including ore deposits in these categories. The fluids are/were aqueous fluids with variable amounts of salts in solution, from measurements on samples in the former cases and from fluid inclusion evidence in the latter cases.

Aqueous fluids

Aqueous fluids are usually considered in two rather different ways:

1:   Emphasis on 'molecular' character—the main end-members of interest are H_2O and CO_2, with H_2, O_2 and so on, which may occur in solution with the H_2O and CO_2. In most aqueous fluids (lacking substantial CH_4) $x_{H_2O} + x_{CO_2}$ does not depart significantly from one because the dissolved salts do not have a substantial diluting effect on the H_2O and CO_2. Thus, it is usual to consider metamorphic fluids to be effectively H_2O–CO_2 mixtures, by saying for example $P_{H_2O} + P_{CO_2} = P_{fluid} = P_{total}$ during metamorphism, even though the metamorphic fluid was undoubtedly a brine.

2:   Emphasis on 'ionic' character—here the interest focuses on the activities of the (ions derived from the) salts in the aqueous fluid. This approach has been adopted primarily for low temperature/low pressure systems because of the almost complete lack of data on the properties of salts in solution at high temperatures and pressures.

As a starting point, we can consider the phase diagram for pure H_2O. The boundary AB in figure 5.1 marks the P–T line across which the regular lattice of the solid (ice) breaks down with increasing temperature. The liquid (water) like the solid is formed of close-packed H_2O molecules. This gives rise to the important property of liquids that they are relatively incompressible, i.e. their volumes are not strong functions of temperature and pressure. In the case of H_2O in the liquid field, it is reasonable to use a standard state of pure liquid water at the temperature and pressure of interest (standard state 1).

The boundary AC marks the P–T line across which the close-packed nature of the liquid breaks down with increasing temperature. The absence of close-packing in the gas (water vapour) results in the volume being a strong function of temperature and pressure, increasing with increasing temperature, and decreasing with increasing pressure. At pressures above the critical point, C, the properties of H_2O change smoothly from those of a liquid to those of a gas with increasing temperature. Similarly at temperatures above the critical point, the properties change smoothly from those of a gas to those of a liquid with increasing pressure. At conditions above the critical point, the phase is called a supercritical fluid.

In most surface and near-surface environments we are considering an

aqueous fluid which is basically water, with or without a vapour containing some proportion of H_2O (for example, air). The AC boundary in
figure 5.1 can be displaced to go through any $P-T$ in the liquid field if the
H_2O in the vapour phase is diluted by an appropriate amount. In most
metamorphic and igneous environments we are considering an aqueous
fluid which is basically supercritical H_2O.

There are two convenient standard states that can be used for gases and
supercritical fluids, depending on the nature of the calculations. The
complications arising in the treatment of standard states for end-members

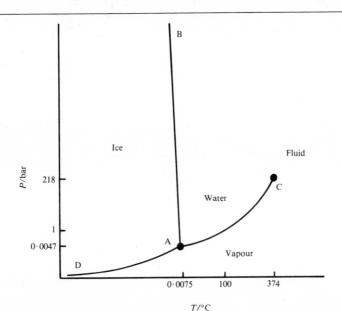

5.1   Schematic phase diagram for H_2O ignoring polymorphic transformations
among the different forms of ice. The temperature and pressure axes are distorted
to clarify the geometry about the intersection, A.

of fluids are covered in appendix C. The appendix is given for complete-
ness and is unnecessary at a first reading.

Comparatively little is known about the $a-x$ relations of fluids at high
temperatures and pressures, particularly for end-members usually present
in very small amounts, for example O_2, H_2, S_2 and so on. In these cases
standard state 4 is most convenient, for example:

$$\mu_{O_2} = G_{O_2}(1, T) + RT \ln\left[x_{O_2} \gamma_{O_2} \exp\left(\frac{\int_1^P V_{O_2}\, dP}{RT} \right) \right] \qquad (5.1)$$

where the exponential term is usually replaced by the fugacity, f, which is
effectively a thermodynamic pressure; in fact, for low pressures and high

temperatures, the fugacity can be replaced by the pressure. Equation (5.1) becomes:

$$\mu_{O_2} = G_{O_2}(1, T) + RT \ln x_{O_2}\gamma_{O_2}f_{O_2} \tag{5.2}$$

As a–x relations in gas mixtures are rarely known, the logarithmic term is not usually considered in terms of its consitituent parts, and the useful equation is:

$$\mu_{O_2} = G_{O_2}(1, T) + RT \ln a_{O_2} \tag{5.3}$$

When using this equation, it must be remembered that although the activity is proportional to the composition of the fluid, it also includes the effect of temperature and pressure on the volume of the pure end-member via the fugacity. Of course, this is one of the advantages of using (5.3) as we do not actually require fugacity tabulations, which are often not available for high temperature and pressure conditions, for example for O_2.

For the major end-members of aqueous fluids, H_2O and CO_2, more is known about the a–x relations at high temperatures and pressures and, further, it is more useful to be able to look at the mole fractions explicitly. In this case either standard state 1 or 4 is convenient. The useful equation for standard state 4 is (5.2). Fugacities of H_2O and CO_2 are tabulated in the literature. Equation (5.2) can be re-arranged to give the useful equation for standard state 1:

$$\mu_{H_2O} = G_{H_2O}(1, T) + RT \ln f_{H_2O} + RT \ln x_{H_2O}\gamma_{H_2O} \tag{5.4}$$

This is the preferred standard state for calculations involving H_2O and CO_2. Tabulations of this fugacity term are required. This term is included in F (appendix A, p 246), and is tabulated in table A.4.

H_2O–CO_2 mixtures show a solvus at low temperatures, for example the consolute temperature is about $500 \, K$ in the range 1 to 3 kbar. The presence of the solvus indicates positive deviations from ideality in these mixtures. A crude regular solution model fit of the available H_2O–CO_2 a–x data gives $w = 20 \cdot 8 - 0 \cdot 015 \, T$ kJ, without a pressure dependence as the H_2O–CO_2 solvus is not obviously pressure dependent, at least above 1 kbar. Thus:

$$RT \ln \gamma_{CO_2} = (20 \cdot 8 - 0 \cdot 015 \, T)x_{H_2O}^2$$
$$RT \ln \gamma_{H_2O} = (20 \cdot 8 - 0 \cdot 015 \, T)x_{CO_2}^2 \tag{5.5}$$

Ions in solution

At least at low temperatures, salts in solution in aqueous fluids are dissociated, they consist of charged ions. For example, NaCl in solution consists of the ions Na^+ and Cl^- (in dilute solutions). Water molecules, although electrically neutral, appear to be quadrupolar to ions adjacent to them because of the distribution of charge in the water molecule. The ions attract the water molecules around them, effectively hydrating the ion. There are also interactions between ions (or the ions plus their hydration spheres) which at relatively low concentrations can be considered electrostatic in nature. At higher concentrations these interactions give rise to the formation of ion-complexes (pairs, triplets, clusters and so on) which may or may not be charged. For example, $CaSO_4$ in solution consists of the ions, Ca^{2+} and SO_4^{2-}, and also the neutral ion-pair $CaSO_4^0$.

The $a-x$ relations of ions in solution are controlled by the structure of the solution, which depends on the concentration of all the ions in solution, the amount these ions are hydrated, the amount of ion-complexing and so on.

Aqueous chemists have not tended to use mole fractions, but an analytical chemistry quantity, either the molality, m, defined as the number of moles of solute species (i.e. ion in solution) per 1000 g of solvent (i.e. H_2O); or the molarity, c, defined as the number of moles of solute species per litre of H_2O. The molality and molarity are both directly proportional to the mole fraction at low concentrations. The molality scale will be used here. For species i, the molality, m_i is:

$$m_i = \frac{1000n_i}{MW_{H_2O}n_{H_2O}} = \frac{55 \cdot 56n_i}{n_{H_2O}}$$

where n_i is the number of moles of i, MW_{H_2O} the molecular weight of water and n_{H_2O} the number of moles of water.

There is a problem in considering the chemical potentials of ions in solution. Consider the $NaCl-H_2O$ system at low temperatures. The $NaCl$ will be completely dissociated and the solution will consist of H_2O with Na^+ and Cl^-, assuming that the H_2O is little dissociated. The solution will be electrically neutral. The definition of the chemical potential of Na^+ is:

$$\mu_{Na^+} = \left(\frac{\partial nG_S}{\partial n_{Na^+}}\right)_{n_{Cl^-}n_{H_2O}}$$

where G_S denotes the Gibbs energy of a mole of solution, the n's refer to the number of moles of each species in the solution, and $n = n_{Na^+} + n_{Cl^-} + n_{H_2O}$. However, this chemical potential has no experimental significance because the definition requires the change of G_S resulting from a change of n_{Na^+} while keeping n_{Cl^-} constant. However this is not possible because the solution will remain electrically neutral. On the other hand, there is no problem with the chemical potential of the electrically neutral 'couple', Na^+Cl^- (with no implication that an ion-pair is formed):

$$\mu_{Na^+Cl^-} = \left(\frac{\partial n G_S}{\partial n_{Na^+Cl^-}} \right)_{n_{H_2O}}$$

where $n_{Na^+Cl^-} = n_{Na^+} + n_{Cl^-}$. This 'couple' is one end-member of our NaCl–H_2O system, and we can revert to our consideration of the chemical potential of end-member 1 in phase A, where 1 is now such a 'couple' and A is an aqueous solution.

A diagram similar to figure 3.3 applies to the chemical potential here, with $\ln x_{1A}$ replaced by $\ln m_{1A}$ (figure 5.2). Thus in the Henry's law region:

$$\mu_{1A} = G_{1A}^* + RT \ln m_{1A}$$

The standard state used here is standard state 2 in terms of molalities rather than mole fractions. This, standard state 2(a), is the hypothetical end-member whose properties are obtained by extrapolating from the Henry's law region to the 1 molal solution at the P–T of interest. Thus:

$$\mu_{1A}^\circ = G_{1A}^* \quad \text{and} \quad a_{1A} = m_{1A} \gamma_{1A}^* \quad \text{with} \quad \gamma_{1A}^* \to 1 \quad \text{as} \quad m_{1A} \to 0$$

$$(5.6)$$

The Henry's law region does not extend to useful concentrations of the ions, the main region of interest is the intermediate region. This means that we must introduce mixing models in order to express the activity coefficients as functions of composition. A bulk composition term for the ions in the solution used in these mixing models is the ionic strength, I, defined as:

$$I = \tfrac{1}{2} \sum m_i z_i^2 \tag{5.7}$$

where z_i is the charge on each ion. There is no contribution to I from neutral ion-complexes. Typical values of I are $0 \cdot 002$ for river water, $0 \cdot 1$ for the water in unlithified sediments, $0 \cdot 7$ for sea water and $7 \cdot 5$ for evaporated sea water about to precipitate halite.

For ionic strengths less than about $0 \cdot 001$, the so called Debye limiting law gives the activity coefficient as:

$$RT \ln \gamma_{1A}^* = A b z_+ z_- \sqrt{I} \tag{5.8}$$

where A is a property of the solvent, and is a function of temperature and pressure, b is the number of ions in the couple (for example 3 in $Na_2^+SO_4^{2-}$), and z_+ and z_- are the charges on the positive and the negative ions in the couple. This equation however is barely adequate even for river water. The next more complicated expression, the Debye–Huckel, is:

$$RT \ln \gamma_{1A}^* = \frac{b A z_+ z_- \sqrt{I}}{1 + a B \sqrt{I}} \tag{5.9}$$

where A and B are properties of the solvent, and are functions of temperature and pressure, while a is an adjustable parameter related to the radii of the ions in the couple. This equation is applicable to solutions with ionic strengths up to $0{\cdot}1$. Even this is not adequate for sea water or brines more saline than this. Both these equations for the activity coefficients can be derived by assuming that all the interactions in the solution are of an electrostatic nature. At higher ionic strengths this assumption is

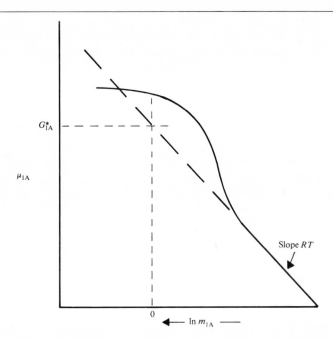

5.2   A plot of μ_{1A} against $\ln m_{1A}$ for an ion 'couple' in solution illustrating standard state 2. For unit molality, $\ln m_{1A} = 0$. At high values of m_{1A} the activity coefficient goes from being less than 1 to greater than 1 (or $RT \ln \gamma_{1A}^{*}$ changes sign).

no longer applicable so that (5.8) or (5.9) cannot be used. Various empirical mixing models have been devised to model activity coefficients for high ionic strength solutions.

An added problem is that the amount of ion-complexing has to be calculated (or measured) before the activities of end-members can be calculated. For example, an analysis of a water may contain so many ppm of SO_4. However not all this SO_4 occurs as SO_4^{2-} ions as some occurs in ion-pairs, for example $CaSO_4^{0}$. The first stage in calculating activities for ions in solution is the calculation of which ions are actually present in the solution. There are some generalizations about the nature and amount of complexing. The higher the charge on the ion, the more likely it is to form abundant complexes. No appreciable complexing occurs between Cl^- and the monovalent and divalent ions at sedimentary temperatures— but even these ions complex strongly at higher (metamorphic) temperatures. Neutral organic molecules form strong complexes with transition metal ions (e.g. Fe^{2+}, Mn^{2+}, etc.). The higher the ionic strength, the more complexing occurs. At high ionic strengths, where complexing is very important, a strongly empirical approach for the calculation of activities is often used. For example, experimentally determined properties are re- lated to an extended form of (5.9), so that the effects of ion-complexing are presumed to be included in the activities, so that the calculation of the amount of complexing is not necessary. Thus the activity of $Ca^{2+}SO_4^{2-}$ would be calculated from the analysed amount of Ca and SO_4 without calculating how much of this Ca is actually present as Ca^{2+} and how much of the SO_4 is present as SO_4^{2-}.

An alternative to using the electrically neutral 'couple' produced by combining ions, for example $Na_2^+SO_4^{2-}$, is to consider a couple with a total number of ions of one, for example $Na_{2/3}^+(SO_4^{2-})_{1/3}$. The properties for such a couple are called mean properties, and are usually denoted by the subscript '±'. The mean activity coefficients are given by (5.8) and (5.9) for the appropriate range of ionic strength, with $b = 1$ as there is a total of one ion in such a couple.

Although there would be no problem in listing G^* for the couples, the alternative is to split G^* for the couple into parts for each of the ions present, for example:

$$G^*_{Na_2^\frac{1}{2}SO_4^{2-}} = 2G^*_{Na^+} + G^*_{SO_4^{2-}}$$

This is done with the help of a reference state for H^+, $G^*_{H^+} = 0$. Thus $G^*_{Cl^-} = G^*_{H^+Cl^-}$, $G^*_{Ca^{2+}} = G^*_{Ca^{2+}Cl_2^-} - 2G^*_{H^+Cl^-}$, and so on. The activity coeffi- cients for individual ions (from which (5.8) and (5.9) are derived) are given by:

$$RT \ln \gamma_i^* = Az_i^2\sqrt{I}$$

or:

$$RT \ln \gamma_i^* = \frac{Az_i^2\sqrt{I}}{1 + aB\sqrt{I}}$$

(5.10)

The value of a for each ion is found by comparing the equation with measurements of the activity coefficient.

One alternative to worrying about activity coefficients is to use the activity without considering this in terms of the molality and the activity coefficient. This approach is particularly relevant in considering an originally-present fluid phase because the activity of the species is what controlled/was controlled by the assemblage, so it is not important to be able to relate the activity to the molality of the species in the fluid. If this is done, it is sometimes convenient to include the volume integral for the ion in the activity in an analogous way to (5.3), because the molar volumes of ions in solution are imperfectly known particularly at high temperatures and pressures. This would be a combination of standard states 2(a) and 4.

The activity of H^+ (in the standard state 2(a)) has been blessed with its own symbol, pH—the definition is $pH = -\log a_{H^+}$. Unfortunately, this is a base 10 logarithm. Converting to the usual base e logarithm:

$$pH = -\frac{1}{2 \cdot 3026} \ln a_{H^+} \quad \text{or} \quad RT \ln a_{H^+} = -3 \cdot 611 \times 10^{-3} T . \, pH \text{ kJ}$$

It is usually more convenient to use a_{H^+} directly in calculations. a_{H^+} is a measure of the acidity of a solution, it is $0 \cdot 0478$ for a neutral solution at $25 \, ^\circ C$ ($pH = 7$), greater than this for an alkaline solution, less than this for an acid solution.

As reactions have to be electrically balanced, some reactions involving ions in different oxidation states have to be balanced using electrons (e) in the solution. For example, for the magnetite (MAG), hematite (HEM), fluid (F) equilibria:

$$\underset{\text{MAG}}{2Fe_3O_4} + \underset{\text{F}}{H_2O} = \underset{\text{HEM}}{3Fe_2O_3} + \underset{\text{F}}{2H^+} + \underset{\text{F}}{2e} \tag{5.11}$$

A reference state of $\mu_e^\circ = 0$ is used for electrons in solution, so $\mu_e = RT \ln a_e$, with no attempt made to express the a_e in terms of the molality of electrons in the solution and an appropriate activity coefficient. The activity of the electron, a_e, is indirectly measurable for a fluid, using electrochemical cells. The oxidation potential, Eh in volts, which can be measured by setting the system up as a half-cell with a reference half-cell, is:

$$Eh = -\frac{RT}{F} \ln a_e,$$

where F is the Faraday constant, $96 \cdot 48$ kJ V^{-1}, which converts volts to kJ. An alternative to Eh, analogous to pH, is pe which is defined as:

$$pe = -\frac{1}{2 \cdot 3026} \ln a_e$$

Therefore $RT \ln a_e = 0 \cdot 01914 T . \, pe = -F . \, Eh$. It is more convenient to use a_e than Eh or pe in calculations. The importance of a_e as a variable, as indicated by the name of Eh, oxidation potential, is that its magnitude is a measure of the oxidation state of the system, the larger is a_e, the more oxidizing the system. For example, in (5.11), hematite is the higher a_e phase, magnetite the lower a_e phase. Sometimes data tabulations for reactions involving electrons are given in terms of potentials in volts.

These can be converted to $\Delta G°$ by multiplying by nF, where n is the number of electrons in the reaction.

Silicate liquids

The most important phase in igneous petrology, the magma, is also the most complicated and least understood phase thermodynamically. The problems arise because the $a-x$ relations of end-members of silicate liquids depend on the internal constitution of the liquid—in the same way as $a-x$ relations in crystalline solids depend on short-range order and site occupancies, and the $a-x$ relations of ions in solution depend on ion-complexing.

Whereas each group of crystalline silicates usually contains one combination of Si–O units (isolated SiO_4 polyhedra in olivines, Si_2O_6 chains in pyroxenes, double chains in amphiboles, sheets in micas, etc.), each silicate liquid contains a great variety of Si–O polymeric units (complexes), which include chains of every length, rings, and so on. One of the major complicating factors in writing $a-x$ relations for end-members of silicate liquids is that they depend on the size and shape distribution of these polymeric units. Further, the environment of each cation (a 'site') depends on the configuration of the polymeric units surrounding it, in contrast to the situation in crystalline silicates where the cations are located in more or less fixed positions and environments (sites) in the oxygen framework. The number and nature of the cation 'sites' in the silicate liquid also affects the $a-x$ relations of end-members of the silicate liquids. Certain cations, particularly Al, are distributed between the cation 'site' in the polymeric units (where the Al substitutes for Si), and the 'sites' interstitial to the polymeric units. Site distributions are just as important in the thermodynamics of silicate liquids as they are in the thermodynamics of crystalline silicates.

The problem is that we have little quantitative information about the size and shape distribution of the polymeric units, the number and nature of the 'sites', or the distribution of elements between these 'sites' in simple silicate liquids, let alone in multicomponent silicate liquids (magmas). However, qualitative conclusions can be reached about the structure of silicate liquids and the relation of structure to $a-x$ relations.

Silica liquid is more or less completely polymerized, the liquid consisting of a SiO_2 framework. The great majority of oxygens are bonded to two silicons. These oxygens have no residual charge and are usually referred to as O^0 oxygens. At breaks in the network, oxygens are bonded to only one silicon. There is unit residual charge on these oxygens and they are usually referred to as O^- oxygens. The polymerized nature of silica liquid accounts for its very high viscosity. On the other hand, many divalent and monovalent oxide liquids (of, for example, CaO and Na_2O) are more or less depolymerized and have low viscosities. In this case the dominant oxygen 'species' is doubly charged oxygen, O^{2-}. In a silicate liquid, the amount of polymerization, and therefore the amounts of O^0, O^- and O^{2-}, will depend on the composition of the liquid and the nature of the cations involved.

A simplistic way of thinking about $a-x$ relations in silicate liquids is in terms of mixing of these three types of oxygen on an oxygen 'site' (O), of mixing cations on a 'site' interstitial to the polymeric units (I), and of mixing silicon and other cations on the cation 'site' within the polymeric units (P). Using an ideal mixing on sites model:

$$a_{SiO_2,L} = x_{Si,P} x_{O^0,O}^2, \qquad a_{CaO,L} = x_{Ca,I} x_{O^{2-},O}, \text{ etc.}$$

which assumes a completely polymerized standard state for SiO_2 and a completely depolymerized standard state for CaO. If the actual structure of the pure oxide does not correspond to the degree of polymerization used in the standard state, then the properties of the actual liquid oxide must be corrected so as to apply to the (hypothetical) standard state oxide. For CaO this correction term would be $-RT \ln x_{Ca,I}^{\circ} x_{O^{2-},O}^{\circ}$, where the $^{\circ}$ superscript means that the mole fractions refer to the actual liquid oxide.

Now, although it is not possible to calculate the mole fractions of the oxygen 'species', we can relate the changes of structure (i.e. polymerization) with composition to changes of activities with composition via the way the mole fractions of the oxygen 'species' change with polymerization. In this, we will concentrate on a_{SiO_2} because a_{SiO_2} is an important parameter in discussing the crystallization of magmas. For example, the crystallization of olivine and orthopyroxene from a magma can be considered in terms of the equilibrium relation for the reaction:

$$Mg_2SiO_4 + SiO_2 = 2MgSiO_3$$
olivine   magma orthopyroxene

Higher a_{SiO_2} will favour the crystallization of orthopyroxene, lower a_{SiO_2} the crystallization of olivine. In crystal fractionation, the way the magma changes composition will therefore depend on a_{SiO_2}, as the effect of crystallizing olivine will be quite different from the effect of crystallizing orthopyroxene. In contrast to orthopyroxene fractionation, olivine fractionation has the important effect of driving basaltic magma compositions towards silica saturation.

If we start with a pure silica liquid, the effect of adding a divalent oxide, MO, will be to break down the SiO_2 framework, and so reduce the proportion of O^0. The effectiveness of this depends on the cation. For example, K_2O is more effective than Na_2O, which is more effective than CaO, which is more effective than FeO, and so on. The more effective the cation is at breaking down the framework, the more rapidly O^0 and therefore a_{SiO_2} decreases as that cation is added to the liquid. Similarly, the a_{SiO_2} of a liquid with a particular proportion of SiO_2 will become lower as the proportion of more effective 'framework-breakers' increases. For example, two similar silicate liquids which differ only in the proportions of Na_2O and K_2O might crystallize orthopyroxene only and olivine only, if the latter liquid has the greater proportion of K_2O.

The effect of the addition of Al_2O_3 on the structure of a silicate liquid depends on the distribution of the Al between the I and P sites, which depends in turn on the other cations present and the proportion of SiO_2 in the liquid. Al acts as a framework-breaker (entering the I site) if there is only a small proportion of I-site cations, but has the opposite effect, entering the P site, when abundant I-site cations are available. This is because the P-site substitution is only energetically favourable if the residual negative charge on an Al which is tetrahedrally coordinated to oxygen in a P site can be locally charge-balanced by a cation, for example Ca or Na. This logic is probably most applicable to more SiO_2-rich liquids. Thus the effect on the a_{SiO_2} of adding Al_2O_3 to a silicate liquid depends on the overall composition of the silicate liquid.

Similarly, complex results can result from the solution of H_2O into a silicate liquid. In more SiO_2-rich liquids, H_2O acts as an efficient 'framework-breaker', reacting with O^0 to enter the breaks in the polymers as OH^- groups. This is the familiar action of H_2O on silicate liquids, resulting in the lowering of a_{SiO_2}. Under other conditions H_2O may enter the liquid simply as a diluent (as H_2O molecules) or by reacting with O^{2-} to form OH^- groups. CO_2 can also enter silicate liquids in different ways with different effects under different conditions.

Although we are a long way from being able to take a silicate liquid composition and write down the activities of the end-members, there is one approach which we can use to considerable effect. This involves defining the activities of end-members in the silicate liquid from the compositions of the minerals in equilibrium with the liquid. For example, we can define the a_{SiO_2} in an olivine–orthopyroxene-bearing magma using the equilibrium relation for the reaction:

$$Mg_2SiO_4 + SiO_2 = 2MgSiO_3$$
olivine   magma orthopyroxene

Here, we need not attempt to relate this a_{SiO_2} to the composition of the magma. Further, we need not have a liquid standard state if this is not convenient. This is the case, for example, for MgO, whose melting temperature is much higher than normal magmatic temperatures. Using a liquid standard state for MgO would be most inconvenient because it would require a long extrapolation of the measured properties of liquid MgO, which obviously can only be collected above the melting temperature.

The main uses of activities defined in this way are:

1: For comparative/descriptive purposes. For example, a_{SiO_2} calculated for different magma types can be related to differences in composition and mineralogy. For example, water activities of volcanics can be related to eruption mechanism. (One conceptually useful way of thinking about the activity of water in magmas, is that the activity is equal to the mole fraction of H_2O in a hypothetical ideal fluid mixture which would be in equilibrium with the magma at the conditions of interest (using a pure H_2O at P–T standard state). This mole fraction can be multiplied by the total pressure to give a water pressure, which therefore also refers to the hypothetical fluid.)

2: For predictive purposes. Knowing, for example, the a_{SiO_2} of a lava at the Earth's surface, the conditions of equilibrium of that magma with other assemblages which also define a_{SiO_2} (phenocrysts, cumulate plutonic nodules, the mantle, etc.) can be calculated. This is Carmichael's a_{SiO_2} method. It can be applied to other end-members of the silicate liquid as long as the activity of the end-member can be calculated from the various assemblages of interest.

Worked examples 5

5(a) Redo worked example 3(f) assuming that the fluid consists of H_2O and CO_2 using (5.5) for the activity coefficient of H_2O.

The equilibrium relation for the assemblage at 2 kbar is now:

$$0 = 238\cdot8 - 0\cdot261T + RT \ln x_{H_2O}^3 + 3(20\cdot8 - 0\cdot015T)x_{CO_2}^2$$

Substituting $R = 0\cdot0083144$ kJ K^{-1} and $x_{CO_2} = 1 - x_{H_2O}$ and simplifying:

$$T = \frac{9570 + 2500(1 - x_{H_2O})^2}{10\cdot5 - \ln x_{H_2O} + 1\cdot8(1 - x_{H_2O})^2}$$

The resulting T–x curve at 2 kbar is plotted in figure 5.3. Including the non-ideality in the fluid has the effect of flattening the curve, so that at low x_{H_2O}, the reaction occurs at higher temperatures. From a geothermometric point of view, changing the metamorphic fluid composition from $x_{H_2O} = 1$ to $0\cdot3$ by dilution with CO_2 has the effect of lowering the equilibrium temperature by only 50 K. This is heartening because one of the worries of performing calculations on non-carbonate metamorphic rocks is that x_{H_2O} is not equal to one because of the presence of CO_2. We must expect substantial CO_2 to have been present in carbonate-bearing metamorphic rocks, and it is sometimes possible to calculate how much by using reactions involving carbonates.

The flattening effect which results from the inclusion of the non-ideality of H_2O–CO_2 mixtures occurs in the other types of reactions in figure 3.8.

For reactions which involve both H_2O and CO_2, the flattening occurs in the central regions of the curves.

5(b)   Plot the position of the hematite (HEM), magnetite (MAG), O_2 (F) reaction, $6HEM = 4MAG + F$,   from $\Delta G^\circ = 491 \cdot 6 - 0 \cdot 3054T$ $-0 \cdot 366P$ kJ (T in K and P in kbar), using standard state 4 for O_2, and assuming that MAG is pure Fe_3O_4 and HEM is pure Fe_2O_3.

The equilibrium relation for this reaction is:

$$\Delta \mu = 0 = 491 \cdot 6 - 0 \cdot 3054T - 0 \cdot 366P + RT \ln \frac{a_{Fe_3O_4,MAG}^4 \, a_{O_2,F}}{a_{Fe_2O_3,HEM}^6}$$

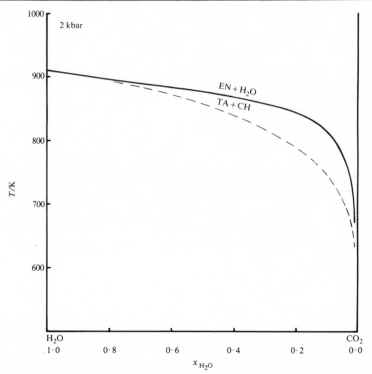

5.3   The T–x diagram for the reaction $TA + CH = EN + H_2O$ at 2 kbar, comparing the curve for H_2O–CO_2 being considered as an ideal solution (broken line, see figure 3.8), with H_2O–CO_2 being considered as a regular solution (5.5). (WE 5(a))

The activities of the solid end-members in the reaction are equal to one, as we are using standard state 1 for these. Thus the equilibrium relation becomes:

$$RT \ln a_{O_2,F} = -491 \cdot 6 + 0 \cdot 3054T + 0 \cdot 366P$$

There are several ways of plotting this equation not least because there are three variables P, T and a_{O_2}. Because block diagrams are not usually very clear, a section through the block diagram is usually plotted. The most familiar is a constant pressure section in which temperature is plotted against activity of oxygen, although constant activity of oxygen and constant temperature plots might be more useful under some circumstances.

At 2 kbar:

$$RT \ln a_{O_2,F} = -490\cdot9 + 0\cdot3054T$$

There are several choices of temperature–oxygen activity plots, but one of the simplest to plot is $\ln a_{O_2}$ against $1/T$ because reaction lines are straight if ΔG° for the reaction is linear in temperature as here. Thus:

$$\ln a_{O_2} = -\frac{59040}{T} + 36\cdot73$$

This is plotted in figure 5.4. We repeat the procedure used in worked example 2(f) to discover which side of the reaction is to higher or lower oxygen activity. At 1200 K and $\ln a_{O_2} = -20$, substituting into the main equilibrium relation above gives $\Delta\mu = -75\cdot2$, which means that the right-hand side of the reaction as written is more stable to lower oxygen activities (i.e. more negative values of $\ln a_{O_2}$). A general rule with these activity plots is that the end-member of interest, in this case oxygen, comes out on the lower activity side of the reaction. It makes sense that the more oxidized assemblage, involving hematite, should be associated with more oxidizing conditions, i.e. involving larger activities of oxygen.

Note that here we are not attempting to relate oxygen activity to amount of oxygen present in the fluid via $a_{O_2} = f_{O_2} x_{O_2} \gamma_{O_2}$. The fugacity could be obtained from tables at least at low pressures, but there are no data for the activity coefficients. The activity coefficient will not be a function of x_{O_2} because we will be in the Henry's law region for oxygen, but the activity coefficient could be quite large, depending on the other end-members making up the fluid. Normally there is little value in knowing the amount of oxygen in the fluid whereas it is interesting to know the activity of O_2.

5(c)   Determine the molality of the species in an aqueous solution at 25 °C and 1 bar which contains 6·01 ppm total Ca, 2·43 ppm total Mg and 24·00 ppm total SO_4 only.

The first step is to convert the ppm values to molalities. The definition of the molality of i is:

$$m_i = \frac{1000}{MW_{H_2O}} \cdot \frac{n_i}{n_{H_2O}}$$

where MW_{H_2O} is the molecular weight of H_2O and the n's refer to the number of moles in the solution. 1 ppm is 10^{-4} wt%. The ppm values must be converted to a mole basis. As we require the ratio of n values, it is sufficient to ratio the wt% for the species and the wt% of H_2O (approximately equal to 100), each divided by the appropriate molecular

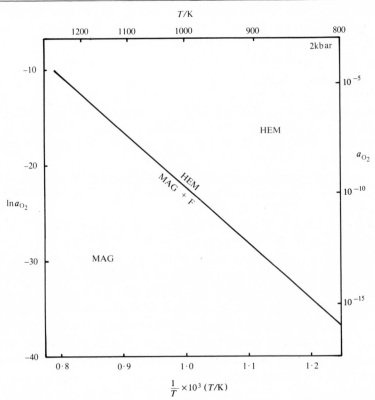

5.4 A $\ln a_{O_2}$–$1/T$ plot showing the position of the HEM = MAG + O_2 reaction. (WE 5(b))

weight. Thus:

$$m_i = \frac{1000}{MW_{H_2O}} \cdot \frac{\dfrac{ppm_i}{MW_i} \times 10^{-4}}{\dfrac{100}{MW_{H_2O}}} = \frac{ppm_i}{MW_i} \times 10^{-3}$$

Using this, the molalities can be calculated from the ppm values. Thus $m_{Ca} = 1.5 \times 10^{-4}$, $m_{Mg} = 1.0 \times 10^{-4}$ and $m_{SO_4} = 2.5 \times 10^{-4}$. These are total molalities because they refer to, for example, the total sulphate in the solution. The next step is to determine how the Ca, Mg and SO_4 are distributed among the species in solution. The main species will be Ca^{2+}, $CaSO_4^0$, Mg^{2+}, $MgSO_4^0$ and SO_4^{2-}. Thus we have three mass balance relations:

$$m_{Ca} = m_{Ca^{2+}} + m_{CaSO_4^0} = 1\cdot5 \times 10^{-4}$$
$$m_{Mg} = m_{Mg^{2+}} + m_{MgSO_4^0} = 1\cdot0 \times 10^{-4}$$
$$m_{SO_4} = m_{SO_4^{2-}} + m_{CaSO_4^0} + m_{MgSO_4^0} = 2\cdot5 \times 10^{-4}$$

We can find out the extent of the formation of these neutral ion-pairs from the equilibrium relations for the reactions:

$$MgSO_4^0 = Mg^{2+} + SO_4^{2-}$$
$$CaSO_4^0 = Ca^{2+} + SO_4^{2-}$$

At 25 °C and 1 bar, $\Delta G° = 13\cdot46$ kJ for the Mg reaction and $\Delta G° = 13\cdot18$ kJ for the Ca reaction, using standard state 2(a) for these species. Substituting R and $T = 298$ K into $\Delta \mu = 0 = \Delta G° + RT \ln K$ for the two reactions and simplifying gives:

$$0\cdot00437 = \frac{a_{Mg^{2+}} a_{SO_4^{2-}}}{a_{MgSO_4^0}} \quad \text{and} \quad 0\cdot00490 = \frac{a_{Ca^{2+}} a_{SO_4^{2-}}}{a_{CaSO_4^0}}$$

The ionic strength has to be calculated in order to find the activity coefficients of the species. The ionic strength, I, is defined as:

$$I = \tfrac{1}{2} \sum z_i^2 m_i$$

where the sum is over all the species in the solution and z_i is the charge on species i. The first problem is that we need to know what we are trying to discover, the molalities of the individual species, in order to calculate I. If it is assumed that the molalities of the neutral ion-pairs are small compared with the other molalities, then the ionic strength can be calculated from the total molalities assuming $m_{Ca^{2+}} = m_{Ca}$ and so on. Thus, z^2 is 4 for each of the species, and:

$$I = \tfrac{1}{2}[4(1\cdot5 \times 10^{-4}) + 4(1\cdot0 \times 10^{-4}) + 4(2\cdot5 \times 10^{-4})] = 0\cdot0010$$

At this low value of the ionic strength, the Debye limiting law should be applicable for the activity coefficients. For this at 25 °C and 1 bar, $A = -2\cdot901$, so that for each divalent species:

$$RT \ln \gamma_i^* = -2\cdot901(4)\sqrt{0\cdot001} = -0\cdot367 \quad \text{or} \quad \gamma_i^* = 0\cdot862$$

The activity coefficients of the neutral ion-pairs should be approximately equal to one at this low ionic strength. Thus substituting into the equilibrium relations:

$$\frac{m_{Mg^{2+}} m_{SO_4^{2-}}}{m_{MgSO_4^0}} = \frac{0\cdot00437}{(0\cdot862)^2} = 0\cdot00588$$

$$\frac{m_{Ca^{2+}} m_{SO_4^0}}{m_{CaSO_4^0}} = \frac{0\cdot00490}{(0\cdot862)^2} = 0\cdot00659$$

The molalities of the five species in the solution can be calculated from these two equations plus the three mass balance relations. This has to be done iteratively, by making an estimate of the answer, and then checking through the equations to obtain a better estimate. This procedure is repeated until the estimate appears to have converged, the estimate not changing through one step of the iteration. This estimate is taken to be the answer.

(1) Start by assuming $m_{MgSO_4^0} = m_{CaSO_4^0} = 0$; then, from mass balance:

$$m_{Mg^{2+}} = 1 \cdot 0 \times 10^{-4}$$

$$m_{Ca^{2+}} = 1 \cdot 5 \times 10^{-4}$$

$$m_{SO_4^{2-}} = 2 \cdot 5 \times 10^{-4}$$

Using these values in the equilibrium relations:

$$m_{MgSO_4^0} = 4 \cdot 25 \times 10^{-6} \quad \text{and} \quad m_{CaSO_4^0} = 5 \cdot 69 \times 10^{-6}$$

(2) Now using these last two values in the mass balance relations:

$$m_{Mg^{2+}} = 0 \cdot 958 \times 10^{-4} \qquad m_{Ca^{2+}} = 1 \cdot 443 \times 10^{-4} \qquad m_{SO_4^{2-}} = 2 \cdot 401 \times 10^{-4}$$

Using these values in the equilibrium relations:

$$m_{MgSO_4^0} = 3 \cdot 91 \times 10^{-6} \quad \text{and} \quad m_{CaSO_{4,}^0} = 5 \cdot 26 \times 10^{-6}$$

(3) Repeating:

$$m_{Mg^{2+}} = 0 \cdot 961 \times 10^{-4} \qquad m_{Ca^{2+}} = 1 \cdot 447 \times 10^{-4} \qquad m_{SO_4^{2-}} = 2 \cdot 408 \times 10^{-4}$$

$$m_{MgSO_4^0} = 3 \cdot 94 \times 10^{-6} \quad \text{and} \quad m_{CaSO_4^0} = 5 \cdot 29 \times 10^{-6}$$

(4) Repeating:

$$m_{Mg^{2+}} = 0 \cdot 961 \times 10^{-4} \qquad m_{Ca^{2+}} = 1 \cdot 447 \times 10^{-4} \qquad m_{SO_4^0} = 2 \cdot 408 \times 10^{-4}$$

$$m_{MgSO_4^0} = 3 \cdot 94 \times 10^{-6} \quad \text{and} \quad m_{CaSO_4^0} = 5 \cdot 29 \times 10^{-6}$$

The molalities have converged. The iteration would have to be continued if more precision (i.e. more decimal places) were required. The actual ionic strength is now $0 \cdot 00096$ rather than $0 \cdot 001$, which was used in the above calculations. The ionic strength and the activity coefficients have to be calculated again at each step in the iteration if the amount of complexing is more significant because I affects the activity coefficients. As the ionic strength of the solution increases, the ions become more complexed and the calculation of the species present in the solution becomes more complicated, not least because simple models cannot be used for the activity coefficients. Also the calculations become more complicated when the solution contains many species. It is not surprising that empirical methods which do not involve the calculation of the actual species present have been devised for sea water and more saline brines. These methods use total molalities and activity coefficients which are supposed to take account of non-ideality and complexing.

5(d)   Consider the stability of water at 25 °C and 1 bar in terms of $\ln a_{H^+}$ and $\ln a_e$.

Water in an aqueous solution will always be partly dissociated to H^+ and OH^-. If the aqueous solution is in equilibrium with a vapour, then a reaction between H_2O, H_2, and O_2 can be considered, with the H_2 and O_2 in the vapour and the H_2O in the solution. The stability of H_2O will be limited on the oxidizing side by the equilibrium between the solution (saturated with oxygen) and a pure oxygen vapour, and on the reducing side by the equilibrium between the solution (saturated with hydrogen) and a pure hydrogen vapour.

Consider the reaction, involving the solution, S, and vapour, V:

$$2H_2O = O_2 + 4H^+ + 4e$$
$$SVSS$$

The $\Delta G°$ for this reaction is 474·5 kJ. Substituting $R = 0·0083144$ kJ K^{-1} and $T = 298$ K into the equilibrium relation and simplifying gives:

$$-191·5 = \ln \frac{a_{O_2,V} a_{H^+,S}^4 a_{e,S}^4}{a_{H_2O,S}^2}$$

At the oxidizing stability limit of water, $a_{O_2,V} = 1$ for a pure O_2 vapour and $a_{H_2O,S} = 1$ because the solubility of O_2 is small even at saturation with oxygen, and so:

$$\ln a_{e,S} = -47·9 - \ln a_{H^+,S}$$

This defines a line on a $\ln a_{H^+} - \ln a_e$ diagram (figure 5.5). Note that if a_{O_2} is less than 1, then the line will move up the diagram by $-\frac{1}{4} \ln a_{O_2}$. Note also that the slope of a reaction line on a $\ln a_1 - \ln a_2$ diagram is defined by the reaction coefficients, by:

$$\frac{d \ln a_1}{d \ln a_2} = -\frac{n_2}{n_1}$$

where n_1 and n_2 are the reaction coefficients for end-members 1 and 2. Thus the slope (but not the position) of the line is independent of thermodynamic data. This means that useful $\ln a - \ln a$ diagrams can sometimes be drawn even in the absence of thermodynamic data. The reaction line in the diagram can be labelled using the principle that e will be on the low $\ln a_e$ side of the reaction.

Consider the reaction:

$$H_2 = 2H^+ + 2e$$
$$VSS$$

For this reaction $\Delta G° = 0$ for reference state reasons. Substituting R and T into the equilibrium relation and simplifying gives:

$$0 = \ln \frac{a_{H^+,S}^2 a_{e,S}^2}{a_{H_2,V}}$$

At the reducing stability limit of water, $a_{H_2,V} = 1$ for a pure H_2 vapour, and so:

$$\ln a_{e,S} = -\ln a_{H^+,S}$$

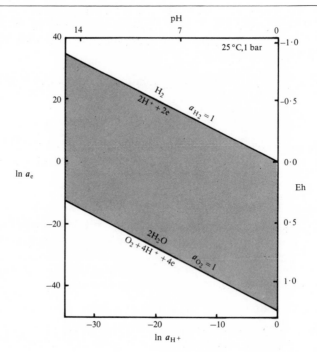

5.5 A $\ln a_e$–$\ln a_{H^+}$ plot showing the stability of H_2O (shaded). The corresponding values of Eh and pH are given on the other axes. (WE 5(d))

This defines another line on the $\ln a$–$\ln a$ diagram (figure 5.5). Note that if a_{H_2} is less than 1, then the line will move down the diagram by $\frac{1}{2}\ln a_{H_2}$.

The shaded area on the diagram puts effective limits on the conditions found in nature, at or near the Earth's surface. The positions of the boundaries will change with temperature and pressure, depending on the P–T dependence of the appropriate $\Delta G°$ values. Ln a–$\ln a$ diagrams like this (and T–$\ln a$ diagrams) follow the same geometric principles as P–T diagrams, particularly in following Schreinemaker's rule (p. 26). Sets of reaction lines can be plotted on $\ln a$–$\ln a$ diagrams, bounding fields where particular assemblages are stable. The water stability lines are usually included on $\ln a_{H^+}$–$\ln a_e$ diagrams to indicate the region which is usually geologically accessible.

5(e) Consider the equilibria involving hematite (HEM), magnetite (MAG) and solution (S) at 25 °C and 1 bar.

There are several ways of doing this! First, we have to constrain the composition of the system of interest. Here we are going to assume that there is no carbonate, sulphide or silicate species in the solution, so that we will not need to consider the relative stabilities of carbonates, sulphides and silicates with the oxides. In this case, a detailed study shows that the major iron ion over most of the range of conditions of interest, the shaded area on figure 5.5, is Fe^{2+}. Thus we will concentrate on the relations between HEM, MAG, Fe^{2+}, H^+ and e.

The relative stabilities of magnetite and hematite can be represented on a $\ln a_{H^+}$–$\ln a_e$ diagram. The reaction used in worked example 4(b):

$$2Fe_3O_4 + \tfrac{1}{2}O_2 = 3Fe_2O_3$$
$$\text{MAG} \quad\;\; \text{V} \quad\quad \text{HEM}$$

can be combined with:

$$H_2O = 2H^+ + 2e + \tfrac{1}{2}O_2$$
$$\text{S} \quad\;\; \text{S} \quad\;\; \text{S} \quad\;\; \text{V}$$

to remove oxygen in the vapour from the resulting reaction:

$$2Fe_3O_4 + H_2O = 3Fe_2O_3 + 2H^+ + 2e$$
$$\text{MAG} \quad\;\; \text{S} \quad\quad \text{HEM} \quad\;\; \text{S} \quad\;\; \text{S}$$

The $\Delta G°$ for this reaction is 42·7 kJ. Substituting this, and R and T into the equilibrium relation for this reaction and simplifying:

$$0 = 17·2 + \ln a_{H^+}^2 a_e^2$$

assuming that the magnetite is pure Fe_3O_4, the hematite is pure Fe_2O_3 and that $a_{H_2O,S}$ is unity. Re-arranging:

$$\ln a_{e,S} = -8·6 - \ln a_{H^+,S}$$

This is plotted in figure 5.6, and the two areas on the diagram are labelled with the stable oxide.

The activity of Fe^{2+} in the solution in the hematite field can be determined by considering the reaction:

$$2Fe^{2+} + 3H_2O = Fe_2O_3 + 6H^+ + 2e$$
$$\text{S} \quad\;\; \text{S} \quad\quad \text{HEM} \quad\;\; \text{S} \quad\;\; \text{S}$$

The $\Delta G°$ for this reaction is 140·3 kJ. Substituting this, and R and T into the equilibrium relation for this reaction and simplifying:

$$0 = 56·64 + \ln \frac{a_{H^+,S}^6 \, a_{e,S}^2}{a_{Fe^{2+},S}^2}$$

Re-arranging:

$$\ln a_{e,S} = -28.3 + \ln a_{Fe^{2+},S} - 3 \ln a_{H^+,S}$$

This can be plotted in figure 5.6 in the form of $\ln a_{Fe^{2+},S}$ contours. The above procedure can be repeated for the reaction involving magnetite and Fe^{2+}:

$$\ln a_{e,S} = -38.2 + \tfrac{3}{2} \ln a_{Fe^{2+},S} - 4 \ln a_{H^+,S}$$

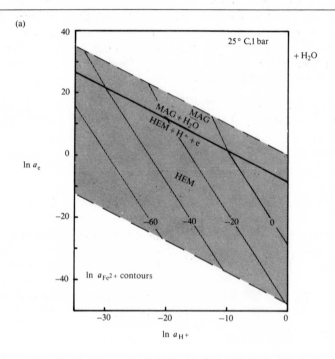

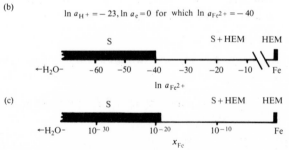

5.6 A $\ln a_e$–$\ln a_{H^+}$ plot, (a), showing the position of the $MAG + H_2O = HEM + H^+ + e$ reaction, and contours of $\ln a_{Fe^{2+}}$ in equilibrium with the appropriate solid phase. (b) and (c) are compatibility diagrams on a $\ln a_{Fe^{2+}}$ and x_{Fe} basis for a particular point on the plot, $\ln a_{H^+} = -23$ and $\ln a_e = 0$, illustrating the relevance of the $\ln a_{Fe^{2+}}$ contours. (WE 5(d))

This can also be plotted in figure 5.6 in the form of contours. Note the inflection of these contours as they pass through the MAG = HEM reaction.

Some idea of what the contours mean can be obtained by looking at compatibility diagrams involving the solution and the appropriate oxide (figures 5.6(b) and (c)). These two compatibility diagrams are for the same values of $\ln a_{H^+}$ and $\ln a_e$ but use different composition scales. The diagram is a projection from oxygen and hydrogen onto the H_2O–Fe line so that Fe_2O_3 plots at Fe. H_2O plots on the line at infinity to the left because logarithmic scales are used. Figure 5.6(b) uses $\ln a_{Fe^{2+}}$ as the axis. The diagram shows that the activity of Fe^{2+} in equilibrium with hematite is -40 and that solutions with lower activities of Fe^{2+} will not be in equilibrium with hematite, at these $\ln a_e$ and $\ln a_{H^+}$ conditions.

This diagram can be transformed in a useful way, if it is assumed that $\gamma^*_{Fe^{2+}} = 1$ at these low concentrations. Then $a_{Fe^{2+}} = m_{Fe^{2+}}$. This can be converted into a mole fraction using $x_{Fe^{2+}} = m_{Fe^{2+}}/55 \cdot 56$. Thus $\ln a_{Fe^{2+}} = -40$ becomes $x_{Fe^{2+}} = 10^{-19 \cdot 1}$. At these $\ln a_{H^+}$ and $\ln a_e$ conditions, figure 5.6(c) is directly comparable to the other compatability diagrams we have considered. For example, a system with $x_{Fe} = 10^{-10}$ will contain HEM and S (of composition $x_{Fe} = 10^{-19 \cdot 1}$); a system with $x_{Fe} = 10^{-30}$ will contain a solution with $x_{Fe} = 10^{-30}$, and so on. Figure 5.6 can be redrawn for particular composition systems. If this is done there will be a field without solid phases which can be labelled with the dominant ion in solution. Thus if the system of interest contains $x_{Fe} = 10^{-19 \cdot 1}$, then the boundary of the magnetite and hematite fields is given by the $\ln a_{Fe^{2+}} = -40$ contour. The field above and to the right of this boundary is the Fe^{2+} field.

Choosing the composition system appropriate to a particular situation involves deciding how much solution is effectively in equilibrium with the solid phase, relating particularly to the mobility of the solution. Clearly the proportion of solids to solution on the ocean floor will be quite different to the proportion in buried sediments. As the sediment moves below the sediment–water interface and becomes compacted, the proportion of fluid decreases substantially, and so the field of Fe^{2+} will decrease substantially moving to higher and higher $\ln a_{Fe^{2+}}$ contours. The processes of dissolution and depositions can be explained in terms of changing the proportions of solution and solids, as well as changing the solution chemistry.

The activities of the other less abundant ions in solution, for example Fe^{3+}, can be calculated in the same way as the activity of Fe^{2+} and can be presented as contours on $\ln a_e$–$\ln a_{H^+}$ diagrams. These diagrams are not the only way of presenting information on the stability and compositions of phases. The problem of deciding which $\ln a_{Fe^{2+}}$ contour to use as the boundary of the hematite and magnetite fields can be avoided by plotting combinations of the activities, $a_e a_{H^+}$ against $a_{Fe^{2+}}/a_{H^+}^2$. This is feasible because reactions can always be constructed so that these combinations account for all the H^+, e and Fe^{2+} in the reaction. The magnetite–hematite equilibrium now becomes:

$$\ln a_e a_{H^+} = -8 \cdot 6$$

The hematite–Fe^{2+} equilibrium:

$$\ln a_e a_{H^+} = -28 \cdot 3 + \ln \frac{a_{Fe^{2+}}}{a_{H^+}^2}$$

The magnetite–Fe^{2+} equilibrium:

$$\ln a_e a_{H^+} = -38 \cdot 2 + \tfrac{3}{2} \ln \frac{a_{Fe^{2+}}}{a_{H^+}^2}$$

The resulting diagram is given in figure 5.7(a). The stability boundaries for water are now horizontal in this plot. Here there is a stability field of Fe^{2+}, regardless of the composition of the system of interest. The disadvantage of this plot is that the activities of other ions cannot be represented on it. However it is possible to construct a block diagram with the extra axis for, say, $a_{Fe^{3+}}/a_{H^+}^3$ (figure 5.7(b)). A new axis is required for each extra ion!

From the above discussions we can see the problems of representing the stability and compositions of the phases of interest, even in such a simple system. When carbonate is added to the system, $a_{CO_3^{2-},s}$ becomes important; as sulphur is added, $a_{SO_4^{2-}}$ becomes important; and so on. From the point of view of tackling geological problems, it is important to identify which parameters will be varying in the system, and which ones can be taken as constant, so that suitable diagrams can be drawn allowing an interpretation of the observed assemblages.

5(f)   What is the pressure–temperature dependence of a_{SiO_2} in magmas generated from the upper mantle? At what conditions did a lava extruded at the Earth's surface last equilibrate with the mantle if it has: (a) $\ln a_{SiO_2} = -1$; (b) $\ln a_{SiO_2} = -1 \cdot 5$ at 1400 K?

The upper mantle consists predominantly of olivine and orthopyroxene, with subsidiary clinopyroxene and garnet (or spinel). A magma in equilibrium with the upper mantle will have its a_{SiO_2} fixed (buffered) by the minerals olivine (OL) and orthopyroxene (OPX), via the thermodynamics of the reaction:

$$\underset{\text{OL}}{Mg_2SiO_4} + \underset{\text{M}}{SiO_2} = \underset{\text{OPX}}{2MgSiO_3}$$

Standard state 3 is used for SiO_2 in the magma by using thermodynamic

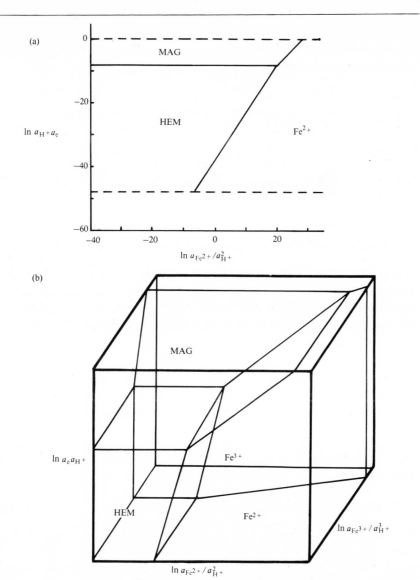

5.7 A $\ln a_e a_{H^+}$–$\ln a_{Fe^{2+}}/a_{H^+}^2$ plot, (a), for MAG–HEM–Fe^{2+} equilibria; and a $\ln a_e a_{H^+}$–$\ln a_{Fe^{2+}}/a_{H^+}^2$–$\ln a_{Fe^{3+}}/a_{H^+}^3$ schematic block diagram, (b), for MAG–HEM–Fe^{2+}–Fe^{3+} equilibria. (WE 5(es))

data for β-quartz for SiO_2. The frequently used alternative is thermodynamic data for SiO_2 glass. For this reaction, $\Delta G° = -14 + 0·008T - 0·36P$ kJ (T in K, P in kbar). The equilibrium relation is:

$$0 = -14 + 0·008T - 0·36P + RT\ln\frac{a^2_{MgSiO_3,OPX}}{a_{Mg_2SiO_4,OL}\,a_{SiO_2,M}}$$

Approximate mineral compositions for the mantle are $x_{Mg,OPX} = 0.85$ and $x_{Mg,OL} = 0.90$. Therefore $a_{MgSiO_3,OPX} = 0.85$ and $a_{Mg_2SiO_4,OL} = 0.81$, assuming ideal mixing for both phases. Substituting these into the equilibrium relation and re-arranging gives:

$$\ln a_{SiO_2,M} = -\frac{1680}{T} + 0.84 - \frac{43P}{T}$$

This allows a P–T diagram to be contoured with $\ln a_{SiO_2,M}$ values for a magma in equilibrium with the mantle (figure 5.8(a)). Thus a magma in equilibrium with the mantle at 60 kbar and 1500 K, either by partial melting at these conditions or by re-equilibrating at these conditions, will have $\ln a_{SiO_2,M} = -2$. Unfortunately this value cannot be related to the composition of the magma at these conditions because we do not know the activity–composition relations for silicate liquids. However useful calculations can be performed if the $\ln a_{SiO_2}$ of a lava extruded at the Earth's surface is known. If the lava contains groundmass olivine and orthopyroxene which crystallized in equilibrium with the magma, the above equilibrium relation can be used to find $\ln a_{SiO_2}$ in the magma at these conditions. Other groundmass assemblages, for example nepheline and alkali feldspar, can be used to fix $\ln a_{SiO_2}$ in the lava at the conditions of crystallization of the groundmass.

Knowing the $\ln a_{SiO_2}$ value for a lava when it is extruded, the pressure and temperature that magma could have been in equilibrium with the mantle can be calculated if we know how $\ln a_{SiO_2}$ evolved during ascent and the pressure–temperature path the magma is likely to have followed on its way to the Earth's surface. If the magma is not fractionated at all from source to surface, the $\ln a_{SiO_2}$ changes by $36P/T$ where P and T refer to the source. This can be included in the $\ln a_{SiO_2}$ expression (figure 5.8(b)):

$$\ln a_{SiO_2} = -\frac{1680}{T} + 0.84 - \frac{79P}{T}$$

The pressure–temperature path of a magma depends on the rate of ascent, size of ascending magma body, convection in the magma body etc. Crudely, the P–T path is likely to lie between $0.3\,°C/km$ (rapid ascent) and $2\,°C/km$ (slow ascent).

A lava with $\ln a_{SiO_2} = -1$ equilibrates with the mantle (intersects the $\ln a_{SiO_2}$ contour on figure 5.8(b) at 12 kbar and 1415 K for rapid ascent, and 13 kbar and 1480 K for slow ascent, giving a calculated depth of origin between 35 and 40 km. For a lava with $\ln a_{SiO_2} = -1.5$, the calculated depth of origin range is 65 to 75 km. If the ascending magma fractionates, the $\ln a_{SiO_2}$ value will change. For example, eclogite fractionation would tend to lower $\ln a_{SiO_2}$. If while the magma is ascending along the $2\,°C/km$, the $\ln a_{SiO_2}$ value is decreased from -1.5 to -2.0 the depth of origin

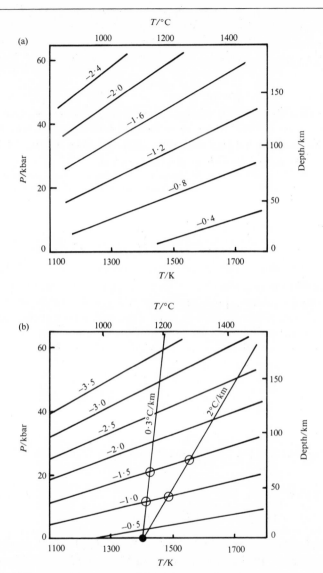

5.8 A $P-T$ diagram, (a), contoured for $\ln a_{SiO_2}$ values for a magma which would be in equilibrium with the mantle (olivine Fo_{90}, orthopyroxene En_{85}) under those conditions. (b) is an equivalent diagram on which the pressure dependence of $\ln a_{SiO_2}$ from source to surface is compensated for. The $\ln a_{SiO_2}$ of a magma at the Earth's surface can be compared directly with the contours to determine the conditions which the magma could have equilibrated with the mantle, given an ascent path (fast ascent, $0.3\,°C/km$; slow ascent, $2\,°C/km$). (WE 5(f))

increases from 75 to 110 km. Thus the calculated depth of origin of a magma is a maximum if the depth of origin is calculated on the basis of no fractionation but fractionation decreasing $\ln a_{SiO_2}$ has occurred. The depth of origin is a minimum if fractionation increasing $\ln a_{SiO_2}$ has occurred.

Obviously the above procedure can be repeated for equilibrium of the magma with other possible source material (the subducted oceanic crust, the lower crust) or with megacrysts or cumulate nodules; also for other end-members of the magma whose activities are well defined in a magma in equilibrium with the mantle or other source.

Problems 5

5(a) The $\Delta G°$ for the reaction:

$$\underset{\text{dolomite}}{3CaMg(CO_3)_2} + \underset{\text{quartz}}{4SiO_2} + \underset{\text{fluid}}{H_2O} = \underset{\text{talc}}{Mg_3Si_4O_{10}(OH)_2} + \underset{\text{calcite}}{3CaCO_3} + \underset{\text{fluid}}{3CO_2}$$

at 2 kbar is:

$$\Delta G° = 173 \cdot 1 - 0 \cdot 2275 T \qquad (T \text{ in } K)$$

using standard state 1 for each end-member in the reaction. Calculate the $T-x$ curve for this reaction at 2 kbar when:
(a) fluid is an ideal solution;
(b) activity coefficients for fluid end-members are given by (5.5);
Assume unit activities for each of the solid end-members and assume that $x_{CO_2} + x_{H_2O} = 1$.

5(b) The $\Delta G°$ for the reaction:

$$\underset{\text{sodalite}}{Na_4Al_3Si_3O_{12}Cl} = \underset{\text{nepheline}}{3NaAlSiO_4} + \underset{\text{fluid}}{NaCl}$$

at 2 kbar is:

$$\Delta G° = 352 \cdot 3 - 0 \cdot 2122 T \qquad (T \text{ in } K)$$

using standard state 1 for sodalite and nepheline and standard state 4 for NaCl. Plot a $\ln a_{NaCl} - 1/T$ diagram for the equilibrium coexistence of pure sodalite and nepheline. At what temperature is nepheline replaced by sodalite during cooling at $\ln a_{NaCl} = -24$?

5(c) Calculate the molalities of Na^+, K^+, $NaSO_4^-$, KSO_4^- and SO_4^{2-} at 25 °C and 1 bar, given that the equilibrium constants for the complexing reactions:

$$KSO_4^- = K^+ + SO_4^{2-} \quad \text{and} \quad NaSO_4^- = Na^+ + SO_4^{2-}$$

are 0·110 and 0·087 respectively at 25 °C, and that the total molalities are $m_{SO_4} = 3 \times 10^{-4}$, $m_{Na} = 2 \times 10^{-4}$ and $m_K = 3 \times 10^{-4}$.

5(d)   Plot a $\ln a_{H^+}$–$\ln a_e$ diagram for equilibria involving native copper (Cu), tenorite (CuO) and cuprite (Cu_2O) at 25 °C and 1 bar and contour it for $\ln a_{Cu^{2+}}$ (for values of -10, -20, -30 and -40) given the Gibbs energy values: $65\cdot0$, 0, $-146\cdot4$, $-127\cdot2$ and $-237\cdot2$ kJ for Cu^{2+}, Cu, Cu_2O, CuO and H_2O respectively.

References

Garrels, R. M. and Christ, C. L., 1965, *Solutions, Minerals and Equilibria*, Harper and Row, London: *and*

Berner, R. A., 1971, *Principles of Chemical Sedimentology*, McGraw-Hill, New York: Further reading in aqueous solution geochemistry.

Bockris, J. O'M and Reddy, A. K. N., 1973, *Modern Electrochemistry 1*, Plenum, New York: Excellent advanced reading on aqueous chemistry.

Holloway, J. R., 1977, Fugacity and activity of molecular species in supercritical fluids: *and*

Eugster, H. P., 1977, Compositions and thermodynamics of metamorphic solutions: *and*

Fraser, D. G., 1977, Thermodynamic properties of silicate melts: *and*

Nicholls, J., 1977, The activities of components in natural silicate melts. In D. G. Fraser (Ed.) *Thermodynamics in Geology*, Reidel, Dordrecht: Advanced reading on the thermodynamics of fluids.

Chapter 6

Equilibrium Thermodynamic Calculations

Thermodynamic calculations are performed using the equilibrium relation for a balanced reaction:

$$\Delta\mu = 0$$

Each chemical potential can be split into two terms, one composition dependent, the other composition independent, thus:

$$\mu_{1A} = G_{1A} + RT \ln x_{1A}\gamma_{1A}$$

where

$$\gamma_{1A} \rightarrow \text{constant} \quad \text{when} \quad x_{1A} \rightarrow 0$$
$$\gamma_{1A} \rightarrow 1 \qquad \text{when} \quad x_{1A} \rightarrow 1$$

This, or re-arrangements of this, can be described in terms of a number of standard states, in terms of:

$$\mu_{1A} = \mu_{1A}^{\circ} + RT \ln a_{1A}$$

The equilibrium relation for a balanced reaction is therefore:

$$\Delta\mu = 0 = \Delta\mu_{1A}^{\circ} + \Delta RT \ln a_{1A}$$

or, written in the usual way:

$$\Delta\mu = 0 = \Delta G^{\circ} + RT \ln K \tag{6.1}$$

Below is a summary of the standard states used in different situations, with some important generalizations about a–x relations. Where the equation is written in terms of the activity a, this implies that no attempt is made to consider the activity in terms of its component parts, mole fraction, activity coefficient and, for standard state 4, the fugacity. The disadvantage of this approach is that the activity cannot be calculated from an analysis of the phase of interest, nor can the composition of the phase be calculated from the activity. However, apart from being the only way of studying end-members in many cases, the end-members usually considered in this way refer to the fluid phase no longer present in the assemblage of interest, either the aqueous metamorphic fluid or a silicate liquid, so that the restriction is not necessarily serious.

Standard state 1: pure end-member at P–T of interest:

$$\mu_{1A} = G_{1A}(P, T) + RT \ln x_{1A} \gamma_{1A}$$

Used for end-members of solids: x and γ usually calculated by mixing on sites. Thermodynamic data in appendix A, table A.3 for some useful end-members.
Used for CO_2 and H_2O in aqueous fluids: γ given by (5.5)
Used for end-members of silicate liquids (alternative is 4).

Standard state 2(a): hypothetical end-member whose properties are obtained by extrapolating from the Henry's law region to the 1 molal solution at the P–T of interest:

$$\mu_{1A} = G_{1A}^* + RT \ln m_{1A} \gamma_{1A}^*$$

Used for ions in solution at low P–T: γ for very dilute solutions from (5.8), (5.9).

Standard state 3: pure end-member in a structural state other than the phase of interest at the P–T of interest:

$$\mu_{1A} = G_{1B}(P, T) + RT \ln a_{1A}$$

Used for end-members of silicate liquids.

Standard state 4: pure end-member at 1 bar and T of interest:

$$\mu_{1A} = G_{1A}(1, T) + RT \ln a_{1A}$$

Used for end-members of aqueous fluids (not usually H_2O and CO_2).

Standard state 2(a)–4: standard state 2(a) but at 1 bar and the T of interest

$$\mu_{1A} = G_{1A}(1, T) + RT \ln a_{1A}$$

Used for ions in solution at high P–T.

We can now summarize the uses of (6.1). Consider a mineral assemblage for which analyses of the phases are available so that activities can be calculated.

Solid–solid reactions: for example

$$NaAlSi_2O_6 + SiO_2 = NaAlSi_3O_8$$
　　　pyroxene　　quartz　　plagioclase

The equilibrium relation for each solid–solid reaction defines a P–T line. Two such reactions with a reasonable angle of intersection will fix both temperature and pressure, while a shallow angle of intersection will result in only poor estimates of temperature and pressure (figure 6.1).

Solid–fluid reactions: If the fluid is still present and can be analysed, then the equilibrium relation for each reaction defines a P–T line. If the fluid phase is absent, the equilibrium relation for each solid–fluid reaction has pressure and temperature as variables, as well as the activities (or amounts) of the end-members of the absent phase which are involved in the reaction. For example for the reaction

$$CaCO_3 + SiO_2 = CaSiO_3 + CO_2$$
　　calcite　　quartz　　woll　　　fluid

the equilibrium relation involves P, T and x_{CO_2} as unknowns. For the reaction

$$H_2O + 3CaMg(CO_3)_2 + 4SiO_2 = Mg_3Si_4O_{10}(OH)_2 + 3CaCO_3 + 3CO_2$$
　　fluid　　　　　dolomite　　　　quartz　　　　　talc　　　　　calcite　　　fluid

the equilibrium relation involves P, T, $x_{CO_2,F}$ and $x_{H_2O,F}$ as unknowns. At least for metamorphic conditions outside the eclogite and granulite facies, P_{fluid} is likely to approach P_{total}, so that $x_{CO_2} + x_{H_2O} = 1$ if CO_2 and H_2O are the most important species in the fluid phase. In this case the unknowns are P, T and x_{CO_2} with $x_{H_2O} = 1 - x_{CO_2}$. Now three reactions with these unknowns are required to define P, T and x_{CO_2} and, again, good angles of intersection are required to fix these values with little error.

For the reaction

$$6H^+ + Mg_3Si_4O_{10}(OH)_2 = 4SiO_2 + 4H_2O + 3Mg^{2+}$$
　　fluid　　　　　　talc　　　　　　quartz　　fluid　　　fluid

the equilibrium relation involves P, T, x_{H_2O}, $a_{Mg^{2+}}$, $a_{H^+}^2$ as unknowns. This number can be decreased by one if $a_{Mg^{2+}}/a_{H^+}^2$ is considered as a variable instead of the activities of the individual ions. If P and T are known then the equilibrium relation fixes a line on a $a_{Mg^{2+}}/a_{H^+}^2$ against x_{H_2O} diagram, and so on.

For the reaction

$$2MgSiO_3 = Mg_2SiO_4 + SiO_2$$

  pyroxene   olivine   magma

the equilibrium relation involves P, T and $a_{SiO_2,magma}$ as unknowns. If, however temperature and pressure are known, then the equilibrium relation fixes $a_{SiO_2,magma}$. Thus the activity of silica in a lava can be calculated if it contains olivine and orthopyroxene and if its crystallization temperature is known.

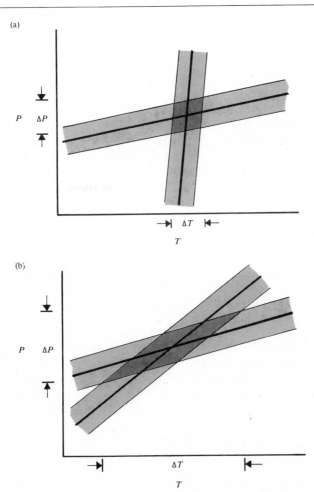

6.1 *P–T* diagrams showing the increase of the uncertainties in the calculated temperature and pressure (given by the heavy shading) as the angle of intersection of the two reaction lines decreases. The uncertainty in the position of each line is shown by light shading.

Thus from a diagrammatic point of view, a line on a P–T diagram contains all the information generated by the equilibrium relation for one solid–solid reaction. However, for a solid–fluid reaction, a P–T diagram can only be contoured for the position of the reaction for different compositions or activities of end-members of the fluid phase. For these reactions, T–x (or T–a), P–x (or P–a) and x–x (or a–a) diagrams can be more useful. We can also adopt some of these types of plots to show how the compositions of minerals change with P or T in particular assemblages. Such diagrams are most important in considering the way the compositions of the phases in an assemblage are likely to have changed as the conditions changed.

Data

The essential data for thermodynamic calculations on mineral assemblages are:

(1) The compositions of the minerals whose end-members are involved in the reactions of interest (plus a knowledge of a–x relations of these minerals) in order to calculate K.

(2) The thermodynamic data for the end-members in the reaction in order to calculate $\Delta G°$ as a function of P–T (for example $\Delta G° = a + bT + cP$).

Some aspects of a–x relations of minerals and fluids of geological interest have been covered in the last two chapters. Some aspects of the calculation of $\Delta G°$ are covered in appendix A, where some useful approximations are outlined. The main useful equation of the appendix is:

$$\Delta G° = \Delta(\Delta_f H(1,298)) - T\Delta S(1,298) + P\Delta V_s(1,298) + nF_{H_2O} + mF_{CO_2}$$

$$(6.2)$$

for a reaction involving n H_2O and $\dot{m}$ CO_2 molecules, using standard state 1 for all the end-members in the reaction. T is in K and P in kbar. $\Delta_f H(1,298)$ is the enthalpy of formation from the elements at 1 bar and 298 K, $S(1,298)$ is the entropy at 1 bar and 298 K, $V(1,298)$ is the volume at 1 bar and 298 K. These values for a set of end-members of interest are given in table A.2. The volume term for each of the fluid end-members (here restricted to H_2O and CO_2) is contained in the F term, which is tabulated in table A.3, in the form:

$$F = a' + b'T$$

covering a particular temperature range at each pressure. Thus ΔV_s refers to the volume change for the reaction of just the solid end-members.

$\Delta G°$ can be considered in the form, $a + bT + cP$, for solid–solid reactions

($n = m = 0$), with:

$$a = \Delta(\Delta_f H(1,298))$$

$$b = -\Delta S(1,298)$$

$$c = \Delta V(1,298)$$

For solid–fluid reactions, it is more awkward because the F terms are involved and they are a complicated function of pressure. However, as they are linear in temperature at each pressure, $\Delta G°$ can be represented as $r + sT$ at each pressure, where:

$$r = \Delta(\Delta_f H(1,298)) + P\Delta V_S(1,298) + na'_{H_2O} + ma'_{CO_2}$$

$$s = -\Delta S(1,298) + nb'_{H_2O} + mb'_{CO_2}$$

which means $r = a + cP$ and $s = b$ for solid–solid reactions.

Equation (6.2) is applicable to solid–solid and solid–fluid reactions (involving H_2O and/or CO_2) as long as r (and s) are much larger than the uncertainties generated by the approximations involved in (6.2). This restriction rules out many solid–solid reactions but few solid–fluid reactions. Thus, for calculations on some solid–solid reactions and most $H_2O \pm CO_2$ bearing reactions, the data in appendix A and (6.2) can be used to calculate $\Delta G°$.

From the point of view of performing thermodynamic calculations, an assemblage or set of assemblages can be considered in terms of some simplified compositional system or model system, the chosen system involving only those end-members for which thermodynamic data are available. The departure of the natural system from the model system is accounted for with the activities of the end-members in the phases in the equilibrium constant for the reaction of interest. What is then of interest is how many independent reactions can be written among the model system end-members in a particular assemblage because this will control how much information (pressure, temperature, fluid phase composition) can be calculated from this assemblage.

Uncertainties

An important practical aspect of calculations concerns uncertainties. Uncertainties in the results of calculations arise because there are uncertainties in the thermodynamic data used to calculate $\Delta G°$ and because there are uncertainties in calculating activities from analyses of minerals. Even if the a–x relations for a mineral are known perfectly, the calculated activity cannot be perfectly known because there are always uncertainties in the chemical analyses of minerals. The best that can be expected in electron microprobe analysis is $\pm 1\%$ of the amount present (referred to as 1% relative) for each oxide.

One way of investigating the uncertainties in a calculation generated by the uncertainties in thermodynamic and analytical data is to repeat the calculation using the data altered by the amount of the uncertainty. This approach becomes tedious if there are several different uncertainties in the data to be considered. First we will examine uncertainties in a qualitative way, then we will use the standard equation for error propagation.

Consider a reaction which we want to use for calculating temperatures at given pressures. Writing $\Delta G° = r + sT$, we have at equilibrium for the reaction:

$$\Delta\mu = 0 = r + sT + RT \ln K$$

Re-arranging, the temperature for a given pressure is:

$$T = \frac{-r}{R \ln K + s} \tag{6.3}$$

The temperature can be calculated from r and s, and K (calculated from the compositions of the minerals). The usual situation is that we know r, s and K imperfectly. The main uncertainty in the calculation of r and s comes from uncertainties in $\Delta_f H(1,298)$ and $S(1,298)$ for the end-members in the reaction. The main uncertainty in the calculation of K comes from uncertainties in the formulation of $a–x$ relations. Considering for the moment that r in (6.3) is well known, then the relative importance of the uncertainties in K and s depends on the relative magnitude of $R \ln K$ and s. If s is much larger than $R \ln K$, then the uncertainty on s will control the uncertainty on the calculated temperature. In this case, the ideal mixing approximation in the calculation of K is unlikely to cause serious errors in the calculated temperature, as the uncertainty on K will have an insignificant effect on the calculated temperature. On the other hand, if $R \ln K$ is much larger than s, the uncertainty in K will control the uncertainty in the calculated temperature. In this case, the ideal mixing approximation is likely to cause serious errors in calculated temperatures. For reactions for which (6.2) is applicable, s tends to be larger than $R \ln K$, unless K is very different from 1, so that uncertainties in K tend to be unimportant. Thus, the ideal mixing approximation can usually be used for calculating K for these reactions.

A major problem in considering uncertainties in thermodynamic calculations is assigning uncertainties to r and s, specifically $\Delta_f H(1,298)$ and $S(1,298)$. The thermodynamic data tabulated in table A.2 in Appendix A are consistent with available experimental work on reactions involving these end-members. In the generation of this table, it was assumed that calorimetric or approximated entropies are correct, and enthalpies of formation were then calculated so that the calculated $P–T$ lines for reactions are consistent with the experimentally determined positions.

Thus, the uncertainties in the $\Delta_f H(1,298)$ and $S(1,298)$ values are related to the uncertainties in the positions of all the reactions used to calculate the thermodynamic data. These uncertainties are not known. A minimum uncertainty on a calculated temperature (from (6.3)) or any other parameter is obtained by assuming that the uncertainty on r is ± 1 kJ and on s is ± 0.001 kJ K^{-1}.

Crude error propagation calculations can be performed using these estimated uncertainties of r and s, and an estimate of the uncertainty of K using the error propagation equation:

$$\sigma_y = \sqrt{\sum \left(\frac{\partial y}{\partial x_i}\right)^2_{x_{j(j\neq i)}} \sigma^2_{x_i}} \quad \text{where} \quad y = f(x_1, x_2, \ldots) \tag{6.4}$$

where σ_y is the uncertainty on y caused by the uncertainties on the x_i, assuming that the uncertainties on the x_i are not correlated. Consider (6.3) where $T = f(r, s, K)$. Now:

$$\left(\frac{\partial T}{\partial r}\right)_{s,K} = \frac{-1}{R \ln K + s} = \frac{T}{r}$$

$$\left(\frac{\partial T}{\partial s}\right)_{r,K} = \frac{r}{(R \ln K + s)^2} = \frac{T^2}{-r}$$

$$\left(\frac{\partial T}{\partial K}\right)_{r,s} = \frac{rR/K}{(R \ln K + s)^2} = \frac{RT^2}{rK}$$

So:

$$\sigma_T = \sqrt{\left(\frac{T}{r}\right)^2 \sigma^2_r + \left(\frac{T^2}{r}\right)^2 \sigma^2_s + \left(\frac{RT^2}{rK}\right)^2 \sigma^2_K} \tag{6.5}$$

Applying this to the muscovite + quartz breakdown reaction (p. 133) for which $r = 71.7$ kJ and $s = -0.0820$ kJ K^{-1} at 2 kbar. If $K = 0.5$, $T = 817$ K, so:

$$\sigma_T = \sqrt{129.8\sigma^2_r + 8.666 \times 10^7 \sigma^2_s + 23964\sigma^2_K}$$

If we take $\sigma_r = 1$ kJ, $\sigma_s = 0.001$ kJ K^{-1} and $\sigma_K = 0.02$, then:

$$\sigma_T = \sqrt{129.8 + 86.60 + 9.59} = 15 \text{ K}$$

Note that this value of σ_K makes an insignificant contribution to the temperature uncertainty. It would have to be 0.09 to increase σ_T to 20 K.

For many dehydration and decarbonation reactions, the suggested uncertainties in r and s cause a minimum uncertainty of about ± 15 K on a calculated temperature at a particular pressure. This means that calculated temperatures should never be quoted to more than the nearest 5 or 10 K.

Equation (6.4) can be applied to any defining equation for a property analogous to (6.3) in order to obtain the uncertainty generated by uncertainties in the thermodynamic and analytical data.

Worked examples 6

Whereas all the worked examples up to this point have involved $\Delta G°$ presented as *faits accompli*, many of the examples here and later will use data presented in appendix A. The first four worked examples here involve looking at some of the previous examples and considering uncertainties in the calculations.

6(a) The best uncertainty on each oxide in an electron probe analysis is about 1% relative. How does this affect the calculation of mole fractions in worked example 4(a)?

The first step is to consider recalculation in an algebraic way, using W_i for the weight % of oxide i, MW_i for the molecular weight of oxide i, NO_i and NM_i for the number of oxygens and number of metals in the formula unit of oxide i, and N for the number of oxygens in the formula unit of the mineral of interest. Using these, the factor F is:

$$F = \frac{N}{\sum_i \dfrac{W_i}{MW_i} NO_i}$$

where the sum is over all the oxides in the analysis. The number of cations per N oxygens for cation j, C_j, is then:

$$C_j = F.NM_j \frac{W_j}{MW_j}$$

There will be a contribution to the uncertainty on C_j from the uncertainties on each of the W_i values via F. Thus

$$\sigma_{C_j}^2 = \sum \left[\left(\frac{\partial C_j}{\partial W_i} \right)_{W_{k(k \neq i)}}^2 \sigma_{W_i}^2 \right]$$

where again the sum is over all the oxides in the analysis.

Considering C_{Ca}, the major sources of uncertainty are in the wt% SiO_2 and CaO; 0·54 and 0·21 respectively for 1% relative of the SiO_2 and CaO in the analysis in worked example 4(a). Thus the uncertainty in C_{Ca} is given by:

$$\sigma_{C_{Ca}}^2 = \left(\frac{\partial C_{Ca}}{\partial W_{SiO_2}} \right)_{W_{k(k \neq SiO_2)}}^2 \sigma_{W_{SiO_2}}^2 + \left(\frac{\partial C_{Ca}}{\partial W_{CaO}} \right)_{W_{k(k \neq CaO)}}^2 \sigma_{W_{CaO}}^2$$

where $C_{Ca} = W_{CaO}F/56 \cdot 08$. Now:

$$\left(\frac{\partial C_{Ca}}{\partial W_{SiO_2}}\right)_{W_{k(k \neq SiO_2)}} = -\frac{W_{CaO}}{56 \cdot 08} \cdot \frac{2}{60 \cdot 09} \cdot \frac{F^2}{N}$$

$$\left(\frac{\partial C_{Ca}}{\partial W_{CaO}}\right)_{W_{k(k \neq CaO)}} = \frac{F}{56 \cdot 08} - \frac{W_{CaO}}{56 \cdot 08} \cdot \frac{1}{56 \cdot 08} \cdot \frac{F^2}{N}$$

Substituting into these:

$$\sigma_{C_{Ca}} = \sqrt{(0 \cdot 00994 \times 0 \cdot 54)^2 + (0 \cdot 03359 \times 0 \cdot 21)^2} = 0 \cdot 0089$$

which is a $1 \cdot 1\%$ uncertainty in the C_{Ca} value of $0 \cdot 8209$, so that one is hardly justified in quoting more than three decimal places for this value. This uncertainty is more or less the same as the uncertainty in CaO (on a relative basis). It is interesting to note that the uncertainty on C_{Ca} is $0 \cdot 0054$ (i.e. $0 \cdot 66\%$) contributed by the uncertainty in SiO_2 even if the analysis of CaO is 'perfect'.

The uncertainty on a mole fraction will be about 1% relative if the uncertainties on the component wt% oxides are 1% relative as long as the mole fraction is constructed directly from C_i values. The uncertainties on the mole fractions of Mg and Fe end-members will depend on the uncertainties in the distribution of Mg and Fe between the two sites, which is largely unknown. The uncertainty in $x_{Al,M1}$ can be calculated using:

$$x_{Al,M1} = C_{Al} - (2 - C_{Si}) = C_{Al} + C_{Si} - 2$$

which is the algebraic way of saying that aluminium fills the tetrahedral site with Si, the remainder being placed on M1. The major contribution to the uncertainty in $x_{Al,M1}$ comes from the uncertainty in W_{SiO_2}. Now:

$$\left(\frac{\partial x_{Al,M1}}{\partial W_{SiO_2}}\right)_{W_{k(k \neq SiO_2)}} = \left(\frac{2W_{Al_2O_3}}{101 \cdot 94} + \frac{W_{SiO_2}}{60 \cdot 09}\right)\left(\frac{-2F^2}{6 \times 60 \cdot 09}\right) + \frac{1}{60 \cdot 09}$$

Substituting into this gives:

$$\sigma_{x_{Al,M1}} = 0 \cdot 00795 \times 0 \cdot 54 = 0 \cdot 0043$$

This is an uncertainty of 14% on the value, $x_{Al,M1} = 0 \cdot 031$. The uncertainty here is much larger because this mole fraction depends on C_{Si} as well as C_{Al}. This means that mole fractions involving $x_{Al,M1}$, and for the same reasons $x_{Al,T}$, will have relatively large errors associated with them. Whether this is acceptable in a calculation will depend on the importance of the equilibrium constant, K, in the equilibrium relation of interest.

6(b) What is the uncertainty in the positions of the water breakdown reactions in worked example 5(d), if the uncertainty on $\Delta G°$ is 2 kJ?

At the oxidizing stability limit of water:

$$\ln a_{e,s} = -\frac{\Delta G^\circ}{4RT} - \ln a_{H^+,s}$$

For a particular value of $\ln a_{H^+,s}$:

$$\sigma^2_{\ln a_{e,s}} = \left(\frac{\partial \ln a_e}{\partial \Delta G^\circ}\right)^2_{\ln a_{H^+,s}} \sigma^2_{\Delta G^\circ}$$

This becomes:

$$\sigma_{\ln a_{e,s}} = \frac{1}{4RT}\sigma_{\Delta G^\circ} = \frac{2}{4RT} = 0\cdot20$$

This is a very small uncertainty in the position of the line in figure 5.5 There is no uncertainty in the position of the other line because $\Delta G^\circ = 0$ for reference state reasons.

6(c) What is the uncertainty in the position of the hematite–magnetite line in worked example 5(e), if the uncertainty in the ΔG° for this reaction is 2 kJ?

Re-arranging the equilibrium relation for this reaction:

$$\ln a_{e,s} = -\frac{\Delta G^\circ}{2RT} - \ln a_{H^+}$$

Continuing as in the last example:

$$\sigma_{\ln a_{e,s}} = \frac{1}{2RT}\sigma_{\Delta G^\circ} = \frac{2}{2RT} = 0\cdot40$$

This, again, is a very small uncertainty in the position of the line.

6(d) What is the uncertainty in the position of the mantle $\ln a_{SiO_2}$ contours in worked example 5(f), if the uncertainty on a is 1 kJ and on b is $0\cdot001$ kJ K^{-1} in $\Delta G^\circ = a + bT + cP$, and the uncertainties on the mantle olivine and orthopyroxene compositions are both $0\cdot05$ on the x_{Mg} value?

The equilibrium expression from which the mantle $\ln a_{SiO_2}$ contours were calculated can be re-arranged to give:

$$P = \frac{1}{c}\left(RT \ln a_{SiO_2} - a - bT - 2RT \ln \frac{x_{Mg,OPX}}{x_{Mg,OL}}\right)$$

The necessary partial differentials from which the uncertainty in the pressure of a contour at a particular temperature can be calculated are:

$$\left(\frac{\partial P}{\partial a}\right)_{b,x_{Mg,OPX},x_{Mg,OL}} = -\frac{1}{c}, \qquad \left(\frac{\partial P}{\partial b}\right)_{a,x_{Mg,OPX},x_{Mg,OL}} = -\frac{T}{c}$$

$$\left(\frac{\partial P}{\partial x_{Mg,OPX}}\right)_{a,b,x_{Mg,OL}} = -\frac{2RT}{c}\frac{1}{x_{Mg,OPX}}$$

$$\left(\frac{\partial P}{\partial x_{Mg,OL}}\right)_{a,b,x_{Mg,OPX}} = \frac{2RT}{c}\frac{1}{x_{Mg,OL}}$$

Substituting into these equations and (6.4), the uncertainty in the position of the contours at 1500 K is 3·3 kbar. This is a reasonable sort of uncertainty and, from the point of view of using diagrams like figure 5.8 to calculate the depth of origin of a lava, this is a small uncertainty compared with the uncertainty due to lack of information on the ascent path of magmas.

6(e) High-grade pelitic rocks often contain quartz (Q), plagioclase (PL), aluminosilicate (AL) and garnet (G). What information concerning the pressure and temperature of formation can be obtained from this assemblage?

The main reactions of interest are the reactions between anorthite, quartz, grossularite and the different aluminosilicates:

$$3CaAl_2Si_2O_8 = Ca_3Al_2Si_3O_{12} + 2Al_2SiO_5 + SiO_2$$

$$\quad\text{PL} \qquad\qquad \text{G} \qquad\qquad \text{AL} \qquad \text{Q}$$

where AL can be any of sillimanite (S), kyanite (K) and andalusite (A). Writing $\Delta G° = a + bT + cP$, then from (2.45) and the data in appendix A, for the reaction involving kyanite:

$$a = \Delta(\Delta_{fH}(1,298)) = 2(-2592) + (-6646) + (-910 \cdot 6) - 3(-4230)$$
$$\quad = -50 \cdot 6 \text{ kJ}$$
$$b = \Delta S(1,298) = 2(0 \cdot 0853) + (0 \cdot 2414) + (0 \cdot 0413) - 3(0 \cdot 2027)$$
$$\quad = -0 \cdot 1548 \text{ kJ K}^{-1}$$
$$c = \Delta V(1,298) = 2(4 \cdot 409) + (12 \cdot 53) + (2 \cdot 269) - 3(10 \cdot 079)$$
$$\quad = -6 \cdot 62 \text{ kJ kbar}^{-1}$$

or:

$$\Delta G_K^° = -50 \cdot 6 + 0 \cdot 1548T - 6 \cdot 62P \text{ kJ} \quad (T \text{ in K, } P \text{ in kbar})$$

with subscript K to denote the kyanite reaction. In a similar way:

$$\Delta G_S^° = -32 \cdot 0 + 0 \cdot 1268T - 5 \cdot 47P$$
$$\Delta G_A^° = -44 \cdot 6 + 0 \cdot 1390T - 5 \cdot 14P$$

The equilibrium relation for the reaction is:

$$\Delta\mu = 0 = \Delta G^\circ + RT \ln \frac{a_{Ca_3Al_2Si_3O_{12},G}\, a^2_{Al_2SiO_5,AL}\, a_{SiO_2,Q}}{a^3_{CaAl_2Si_2O_8,PL}}$$

Quartz is usually pure SiO_2 and aluminosilicate pure Al_2SiO_5, so $a_{SiO_2,Q} = 1$ and $a_{Al_2SiO_5,AL} = 1$. Using ideal mixing on sites for garnet, then $a_{Ca_3Al_2Si_3O_{12},G} = x^3_{Ca,G}$, as the Al and Si are usually only diluted to a small extent on their respective sites. Using ideal molecular mixing for plagioclase, $a_{CaAl_2Si_2O_8,PL} = x_{Ca,PL}$. So the equilibrium relation for the reaction can be written:

$$0 = \Delta G^\circ_{AL} + RT \ln \frac{x^3_{Ca,G}}{x^3_{Ca,PL}} = \Delta G^\circ_{AL} + 3RT \ln \frac{x_{Ca,G}}{x_{Ca,PL}}$$

This equation is a relation between four variables, P, T, $x_{Ca,G}$ and $x_{Ca,PL}$. Usually we know the composition terms (by analysing the compositions of the phases in the assemblage), so the first diagram to draw is a P–T diagram. This can be contoured for values of $K = (x_{Ca,G}/x_{Ca,PL})^3$. For the $K = 0.002$ contour, there are three equilibrium relations corresponding to the reaction written with each of the aluminosilicate polymorphs:

K $0 = -50.6 + 0.1031T - 6.62P$

S $0 = -32.0 + 0.0751T - 5.47P$

A $0 = -44.6 \times 0.0873T - 5.14P$

Each of these defines a straight line. The stable line is the one which involves the stable aluminosilicate, so first the aluminosilicate phase diagram must be drawn. Again, using the data in appendix A, for:

K = S $\Delta\mu = 0 = 9.3 - 0.0140T + 0.575P$

K = A $\Delta\mu = 0 = 3.0 - 0.0079T + 0.738P$

A = S $\Delta\mu = 0 = 6.3 - 0.0061T - 0.163P$

These can be plotted to give the aluminosilicate phase diagram (figure 6.2, see figure 2.11). The $K = 0.002$ contour can now be drawn on the diagram. Other contours can be generated and plotted in the same way.

Thus a rock with the Q–PL–AL–G assemblage, if representing frozen-in equilibria, will have formed on the appropriate K contour. So if an estimate of temperature is already available, say 600 °C for the $K = 0.002$ contour, then the pressure can be read off the diagram as 6 kbar. If the assemblage contained two aluminosilicates, say K and S, at equilibrium, then the intersection of the K = S boundary and the appropriate K contour will fix both temperature and pressure.

The contours on figure 6.2 are quite general, applying to any rock compositions which contain $Q + PL + AL + G$. It is more difficult to consider the compositions of the phases which might occur in particular rock compositions at particular temperatures and pressures. Here we need to consider the equilibrium relation for the reaction plus some mass balance relations. Consider a sequence of rocks in which all the Ca is involved in plagioclase and garnet, and that the plagioclase is an albite–anorthite solid solution and that the garnet is an almandine–grossularite solid solution. The composition of such rocks could be plotted in the tetrahedron, figure 6.3. If there are z_{Ca} atoms of Ca in the rock and N_G and N_{PL} molecules of garnet and plagioclase, then:

$$z_{Ca} = 3x_{Ca,G}N_G + x_{Ca,PL}N_{PL}$$

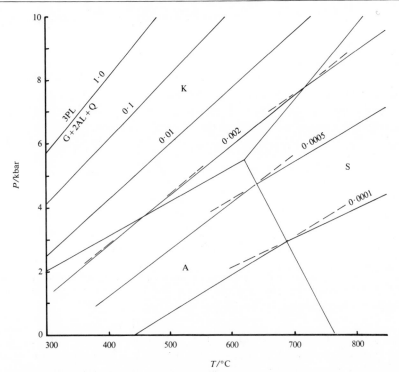

6.2 A P–T diagram showing K contours for the $3PL = G + 2AL + Q$ reaction intersecting the aluminosilicate phase diagram. (WE 6(e))

Similarly for Na and Fe, assuming that they occur only in the plagioclase and garnet respectively:

$$z_{Na} = (1 - x_{Ca,PL})N_{PL} \quad \text{and} \quad z_{Fe} = 3(1 - x_{Ca,G})N_G$$

Re-arranging the last two and substituting into the first:

$$z_{Ca} = \frac{x_{Ca,G}}{1 - x_{Ca,G}} z_{Fe} + \frac{x_{Ca,PL}}{1 - x_{Ca,PL}} z_{Na}$$

Dividing through both sides by $z_{Ca} + z_{Fe} + z_{Na}$, and denoting the proportion of $Ca + Fe + Na$ that is Ca by y_{Ca} (i.e. $y_{Ca} = z_{Ca}/(z_{Ca} + z_{Fe} + z_{Na})$), and re-arranging:

$$x_{Ca,PL} = \frac{y_{Ca} - x_{Ca,G}(y_{Ca} + y_{Fe})}{1 - y_{Fe} - x_{Ca,G}}$$

This relates the composition of the garnet to the composition of the plagioclase via the proportions, y.

Substituting this into the equilibrium relation for the reaction involving kyanite at 823 K:

$$0 = -50 \cdot 6 + 0 \cdot 1548(823) - 6 \cdot 62P + 3R(823) \ln \frac{x_{Ca,G}(1 - y_{Fe} - x_{Ca,G})}{y_{Ca} - x_{Ca,G}(y_{Ca} + y_{Fe})}$$

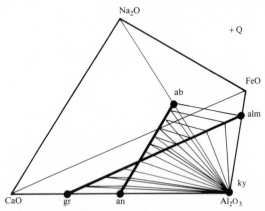

6.3 A tetrahedral compatibility diagram showing the effect of adding Na_2O and FeO to the simple system. The position of the tie lines will be a function of temperature and pressure. (WE 6(e))

This relates the composition of the garnet to pressure for any rock composition expressed in terms of the y values. The composition of the plagioclase at each pressure can then be obtained from the mass balance relation. The compositions of the phases are strongly dependent on rock composition as well as pressure as can be seen in figure 6.4. Note that in this diagram the garnet is less calcium-rich than the feldspar below 11·6 kbar, and more calcium-rich above 11·6 kbar. This can also be noted in figure 6.2, as K is less than 1 below 11·6 kbar at 823 K, whereas above 11·6 kbar it is greater than 1.

This type of diagram allows observation of the way the compositions of the coexisting minerals change with changing temperature or pressure, information that cannot be obtained from $P-T$ diagrams. On the other hand, $P-T$ diagrams can be contoured for $x_{Ca,G}$ and $x_{Ca,PL}$ for particular rock compositions using the above equations, given the assumptions involved in the mass balance relations.

6(f) The breakdown of muscovite (MU) + quartz (Q) to give aluminosilicate (AL) + alkali feldspar (AF) + fluid (F) commonly occurs with progressive metamorphism at the upper end of the amphibolite facies. How does this assemblage constrain the conditions of metamorphism?

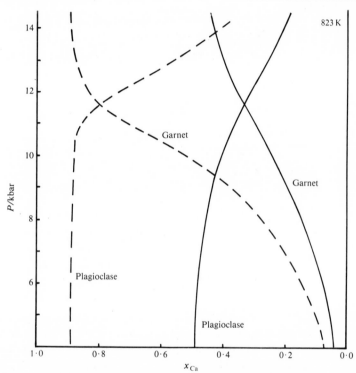

6.4 The compositions of G and PL as a function of pressure for a particular formulation of the mass balance relations for two compositions; see text. (WE 6(e))

Using the thermodynamic data in appendix A for the main reaction of interest:

$$KAl_3Si_3O_{10}(OH)_2 + SiO_2 = Al_2SiO_5 + KAlSi_3O_8 + H_2O$$
$$\quad\;\; MU \qquad\qquad Q \qquad AL \qquad AF \qquad F$$

we get for each aluminosilicate polymorph:

$$\Delta G_A^\circ = 98\cdot0 - 0\cdot1747T - 0\cdot31P + F_{H_2O}$$
$$\Delta G_S^\circ = 104\cdot3 - 0\cdot1808T - 0\cdot48P + F_{H_2O}$$
$$\Delta G_K^\circ = 95\cdot0 - 0\cdot1668T - 1\cdot05P + F_{H_2O} \quad (T \text{ in K, } P \text{ in kbar})$$

where F_{H_2O} can be looked up in the tables in appendix A. Consider that the muscovite consists of pure $KAl_3Si_3O_{10}(OH)_2$, the alkali feldspar of pure $KAlSi_3O_8$, aluminosilicate of pure Al_2SiO_5 and quartz of pure SiO_2. Then the equilibrium relation for this reaction is:

$$\Delta\mu = 0 = \Delta G^\circ_{AL} + F_{H_2O} + RT \ln x_{H_2O,F}$$

if the fluid is an ideal mixture or if P_{fluid} is less than P_{total} (in which case $x_{H_2O} = P_{fluid}/P_{total}$ if the fluid consists primarily of H_2O). If the H_2O in the fluid is diluted by CO_2, then an activity coefficient has to be included, giving:

$$\Delta\mu = 0 = \Delta G^\circ_{AL} + F_{H_2O} + RT \ln x_{H_2O,F} + (20\cdot8 - 0\cdot015\,T)x^2_{CO_2,F}$$

In the same way as in the last worked example, a P–T diagram can be contoured for $x_{H_2O,F}$. Consider the contour for $x_{H_2O} = 1$. From the tables, for 2 kbar and a temperature range of 873–973 K:

$$F_{H_2O} = -25\cdot7 + 0\cdot0927\,T$$

Substituting this, $P = 2$ kbar and $x_{H_2O} = 1$ into the equilibrium relation without the activity coefficient for the reaction involving andalusite:

$$0 = 98\cdot0 - 0\cdot1747\,T - 0\cdot31(2) - 25\cdot7 + 0\cdot0927\,T + RT \ln(1)$$
$$= 71\cdot68 - 0\cdot0820\,T$$

which gives $T = 874$ K $= 601\,°C$. This would be quoted as $600\,°C$ as the 1 does not have any significance with the likely ±15 K on the calculated temperature from uncertainties in the thermodynamic data. Using the wrong temperature range for F_{H_2O} has little effect on the calculated temperature. The calculated temperature is $607\,°C$ if F_{H_2O} for 2 kbar and 1073–1173 K is used.

Both sets of contours in figure 6.5 were calculated using the above procedure. Here the contours are curved because F_{H_2O} is not linear in pressure. Note that the H_2O–CO_2 contours are at a higher temperature than the other contours. A horizontal section through this diagram can be represented as a T–x_{H_2O} diagram, the result being analogous to figure 5.3.

The presence of the MU + Q breakdown reaction in a sequence of metamorphic rocks only constrains the metamorphic conditions if the composition of the metamorphic fluid is known. If x_{H_2O} approaches 1, then the reaction does indeed take place at high temperature. If the H_2O is progressively diluted with CO_2, then the reaction takes place at progressively lower temperatures. The same is true if P_{H_2O} is less than P_{total}. If there are no end-members capable of diluting the H_2O and P_{H_2O} is less than P_{total}, this means that there is no fluid present. The x_{H_2O} is in this case a hypothetical quantity, referring to the composition of an H_2O fluid diluted by some end-member with which it mixes ideally, which would be in equilibrium with the assemblage at those conditions. The lower temperature contours will be metastable with respect to assemblages involving carbonates if the H_2O is diluted by CO_2. The reaction usually takes place in the field at higher grade than the aluminosilicate triple point (about 900 K at 5·5 kbar) when this is identifiable. This restricts x_{H_2O} to be greater than 0·47 for $P_{H_2O} = P_{fluid} < P_{total}$, and to be greater than 0·25 if H_2O is diluted by CO_2.

Figure 6.5 applies to assemblages in which the muscovite is pure $KAl_3Si_3O_{10}(OH)_2$ and the alkali feldspar is pure $KAlSi_3O_8$, whereas natural muscovites contain paragonite ($NaAl_3Si_3O_{10}(OH)_2$) and margarite

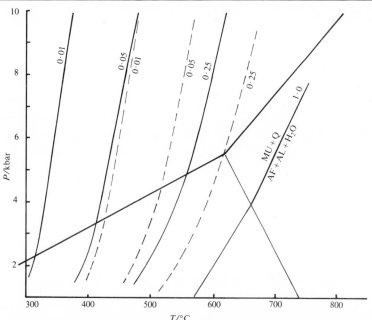

6.5 A P–T diagram showing the position of contours of x_{H_2O} for the MU + Q = AF + AL + H_2O reaction intersecting the aluminosilicate phase diagram. Full lines for H_2O diluted with something with which it mixes ideally; broken lines for H_2O diluted by CO_2. (WE 6(f))

($CaAl_4Si_2O_{10}(OH)_2$) in solid solution, while natural alkali feldspars can contain albite and anorthite in solid solution. In the Na system there is a paragonite + quartz breakdown reaction and in the Ca system a margarite + quartz breakdown reaction. Both these reactions take place at a lower temperature (at a particular pressure and x_{H_2O}) than the muscovite + quartz breakdown reaction. This results in lower temperatures for muscovite solid solution, alkali feldspar solid solution, aluminosilicate, quartz assemblages at a particular pressure and x_{H_2O} in the Na–K–Ca system as can be seen from the calculations on the Na–K system below.

The new reaction of interest is:

$$NaAl_3Si_3O_{10}(OH)_2 + SiO_2 = NaAlSi_3O_8 + Al_2SiO_5 + H_2O$$

$$\quad\quad MU \quad\quad\quad\quad Q \quad\quad\quad AF \quad\quad\quad\quad AL \quad\quad F$$

From the data in appendix A, $\Delta G° = 110·3 - 0·1972T - 0·29P$ involving andalusite as the aluminosilicate. The equilibrium temperature at 2 kbar is $530\,°C$ for the 'pure' assemblage paragonite + albite + andalusite + quartz + H_2O. For equilibrium between $MU + AF + A + Q + H_2O$ at 2 kbar:

$$0 = 71·7 - 0·0820T + RT \ln \frac{a_{KAlSi_3O_8,AF}}{a_{KAl_3Si_3O_{10}(OH)_2,MU}}$$

and:

$$0 = 84·0 - 0·1045T + RT \ln \frac{a_{NaAlSi_3O_8,AF}}{a_{NaAl_3Si_3O_{10}(OH)_2,MU}}$$

If the muscovite and alkali feldspar solid solutions are ideal solutions and binary $(x_{Na} = 1 - x_K)$ then:

$$0 = 71·7 - 0·0820T + RT \ln \frac{x_{K,AF}}{x_{K,MU}}$$

$$0 = 84·0 - 0·1045T + RT \ln \frac{1 - x_{K,AF}}{x_{K,MU}}$$

The equations can be solved at each temperature to give the compositions of the muscovite and alkali feldspar solid solutions. Re-arranging:

$$\exp\left[-\left(\frac{71·7 - 0·0820T}{RT} \right) \right] = A = \frac{x_{K,AF}}{x_{K,MU}}$$

$$\exp\left[-\left(\frac{84·0 - 0·1045T}{RT} \right) \right] = B = \frac{1 - x_{K,AF}}{1 - x_{K,MU}}$$

Re-arranging each again:

$$x_{K,AF} = Ax_{K,MU} \quad \text{and} \quad x_{K,AF} = 1 - B(1 - x_{K,MU})$$

Subtracting one from the other and re-arranging:

$$x_{K,MU} = \frac{1 - B}{A - B}$$

Substituting back:

$$x_{K,AF} = \frac{A(1 - B)}{A - B}$$

The T–x diagram (figure 6.6) is constructed by substituting into these equations at each temperature. Note the similarity between this diagram and figure 2.16. It can be interpreted in exactly the same way as long as all the Na and K in the rock are in the muscovite and alkali feldspar solid solutions. The x-axis can be thought of as K/(K + Na) for the rock as well as giving the compositions of the minerals. Thus, during progressive metamorphism (represented by passage up a vertical line on the diagram), the first alkali feldspar formed would be Na-rich and would become more K-rich as the muscovite becomes more K-rich until it disappears. Note that the MU + AF + A + Q assemblage will be found over a distance in the

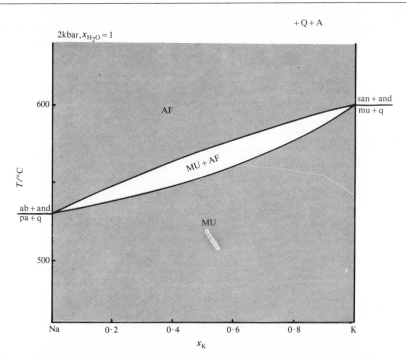

6.6 A $T-x$ diagram for MU–AF–AL–Q–H_2O equilibria at 2 kbar and $x_{H_2O} = 1$, assuming MU and AF form ideal solutions between muscovite and paragonite and between albite and sanidine respectively. (WE 6(f))

field in the prograde direction, this distance depending on the composition of the system, for example a distance equivalent to 12 °C at K/(K + Na) = 0·8 for the rock.

Unfortunately neither muscovite nor alkali feldspar solid solutions are ideal; both are strongly positively non-ideal so that the central part of figure 6.6 is disturbed by two solvi, one between albite and sanidine solid solutions, the other between muscovite and paragonite solid solutions. Nevertheless the terminal regions of the $T-x$ loop are substantially correct and allow the above interpretation of the behaviour of the muscovite + quartz breakdown in the Na–K system. The same logic applies in the ternary K–Na–Ca system.

6(g) A high-grade pelite consists of quartz, alkali feldspar ($Na_{0.1}K_{0.9}Al$-Si_3O_8) plagioclase ($Na_{0.4}Ca_{0.6}Al_{1.6}Si_{2.4}O_8$), garnet ($Fe_{2.86}Ca_{0.14}Al_2Si_3O_{12}$), muscovite ($K_{0.95}Na_{0.05}Al_3Si_3O_{10}(OH)_2$) and sillimanite; it originally contained an H_2O–CO_2 fluid phase. Calculate the conditions of formation assuming each of the solid solutions is ideal. Consider the uncertainties in the calculations.

This is the usual sort of geothermometry/geobarometry problem: given an assemblage, the compositions of the minerals, and presuming equilibrium calculate the conditions of formation of the assemblage. In this case we can use the results of the last two worked examples.

The phases in the muscovite + quartz breakdown reaction are present in this assemblage. Therefore one equilibrium relation is:

$$0 = 104 \cdot 3 - 0 \cdot 1808 T - 0 \cdot 48 P + F_{H_2O} + RT \ln \left(\frac{0 \cdot 90}{0 \cdot 95} x_{H_2O,F} \right)$$

$$+ (20 \cdot 8 - 0 \cdot 015 T) x_{CO_2}^2$$

At 4 kbar this becomes:

$$0 = 76 \cdot 1 - 0 \cdot 0818 T + RT \ln x_{H_2O,F} + (20 \cdot 8 - 0 \cdot 015 T)(1 - x_{H_2O})^2$$

At 6 kbar:

$$0 = 76 \cdot 4 - 0 \cdot 0782 T + RT \ln x_{H_2O,F} + (20 \cdot 8 - 0 \cdot 015 T)(1 - x_{H_2O,F})^2$$

The x_{H_2O} contours on figure 6.7 can be drawn using these two equations.

The phases in the plagioclase–garnet–aluminosilicate–quartz reaction are also present in this assemblage. Therefore another equilibrium relation for this assemblage is:

$$0 = -32 \cdot 0 + 0 \cdot 1268 T - 5 \cdot 47 P + RT \ln \left(\frac{0 \cdot 0466}{0 \cdot 6} \right)^3$$

$$= -32 \cdot 0 + 0 \cdot 0631 T - 5 \cdot 47 P$$

This straight line can also be plotted on figure 6.7. The intersection of this line with the x_{H_2O} contours gives the range of conditions of formation of the assemblage. The limits are:

$x_{H_2O} = 1$ $T = 685\,°C$ and $P = 5 \cdot 2\,kbar$

$x_{H_2O} = 0 \cdot 6$ $T = 635\,°C$ and $P = 4 \cdot 6\,kbar$

The temperature and pressure are well constrained.

The uncertainties will be considered for the intersection of the $x_{H_2O} = 1$ contour for the muscovite reaction and the line for the garnet reaction.

The uncertainties used will be ± 1 kJ in a and ± 0.001 kJ K^{-1} in b, in $\Delta G° = a + bT + cP$. The uncertainty in the position of the muscovite reaction can be calculated using (6.4) and (6.5). At 4 kbar, $r = 76.1$, $s = -0.0818$, $T = 930$ K and $K = 0.9474$. Both the alkali feldspar and muscovite end-members will be near the Raoult's law region so the main uncertainty in the equilibrium constant, K, will be from the analyses of the phases. Thus 2% relative (i.e. 0.019) would be a reasonable uncertainty for K here. Substitution into the error propagation equation gives the uncertainty in T to be 17 K. The uncertainty in K makes a trivial contribution to this. At 6 kbar, the uncertainty is 18 K.

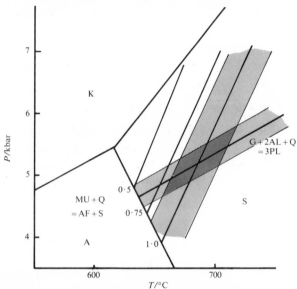

6.7 A $P–T$ diagram for the geothermometry–geobarometry problem, WE 6(g). The steep lines are contours for x_{H_2O} in a $H_2O–CO_2$ fluid for MU + Q + AF + S + F. The shallow line is for G + AL + Q + PL. The minimum uncertainties for each reaction are given by light shading (at $x_{H_2O} = 1$ for the MU + Q reaction), the overlap of which (heavy shading) gives an estimate of the uncertainty for the calculated temperature and pressure (for $x_{H_2O} = 1$ for the MU + Q reaction).

The uncertainties for the other reaction can be considered using the same formula. For a 2% relative error on K, the uncertainty in the temperature at 4 kbar is 21 K, at 6 kbar is 23 K. The intersection of the uncertainty bands for the two reactions gives the uncertainty in the conditions of formation of the assemblage (at $x_{H_2O} = 1$). In fact, the calculated temperature and pressure are likely to be minima because the K value for the garnet reaction is likely to be a minimum. This is because grossular–almandine garnets are probably positively non-ideal at these temperatures and pressures, and grossular could be in the Henry's law region.

This will increase the K value for the reaction. If the activity coefficient for Ca$^{\cdot}$ in the garnet is $1 \cdot 5$, the intersection moves up to $7 \cdot 5$ kbar and $740\,^{\circ}\text{C}$ at $x_{H_2O} = 1$.

6(h) Over what temperature–x_{CO_2} range is tremolite stable in calcite-bearing limestones in the system CaO–MgO–SiO$_2$–CO$_2$–H$_2$O for assemblages involving calcite (CC), dolomite (DOL), tremolite (TR), quartz (Q) and diopside (DI), coexisting with an H$_2$O–CO$_2$ fluid at 4 kbar?

If calcite is present in excess, the most useful compatibility diagram here will be a projection from CC in figure 2.24. The phases of interest here are plotted in figure 6.8(a). There are four reactions of interest, one of the phases is left out in each, [DOL], [TR], [DI] and [Q]. The [CC] reaction is not considered because all the assemblages are considered to have excess calcite. The reactions are:

[DI] $5\text{DOL} + 8\text{Q} + \text{H}_2\text{O} = \text{TR} + 3\text{CC} + 7\text{CO}_2$

[TR] $\text{DOL} + 2\text{Q} = \text{DI} + 2\text{CO}_2$

[DOL] $\text{TR} + 3\text{CC} + 2\text{Q} = 5\text{DI} + 3\text{CO}_2 + \text{H}_2\text{O}$

[Q] $\text{TR} + 3\text{CC} = 4\text{DI} + \text{DOL} + \text{CO}_2 + \text{H}_2\text{O}$

Assuming that the solid phases consist of the appropriate pure end-member, and using the thermodynamic data in appendix A, including F_{H_2O} and F_{CO_2} for 773–873 K and 4 kbar, the equilibrium relations for these reactions are:

for [DI]

$$\Delta\mu = 0 = 464 \cdot 6 - 1 \cdot 0271T - 11 \cdot 5(4) - 1(-27 \cdot 2 + 0 \cdot 1004T)$$
$$+ 7(0 \cdot 6 + 0 \cdot 0832T) + RT \ln K$$
$$= 450 \cdot 0 - 0 \cdot 5451T + RT \ln \left(\frac{x^7}{1-x}\right) + (20 \cdot 8 - 0 \cdot 015T)$$
$$\times [7(1-x)^2 - x^2]$$

using x for x_{CO_2} and $1-x$ for x_{H_2O};

for [TR]

$$\Delta\mu = 0 = 156 \cdot 2 - 0 \cdot 3326T - 4 \cdot 36(4) + 2(0 \cdot 6 + 0 \cdot 0832T) + RT \ln K$$
$$= 140 \cdot 0 - 0 \cdot 1662T + RT \ln x^2 + (20 \cdot 8 - 0 \cdot 015T)2(1-x)^2$$

for [DOL]

$$\Delta\mu = 0 = 316 \cdot 4 - 0 \cdot 6354T - 0 \cdot 86(4) + (-27 \cdot 2 + 0 \cdot 1004T)$$
$$+ 3(0 \cdot 6 + 0 \cdot 0832T) + RT \ln K$$
$$= 251 \cdot 6 - 0 \cdot 2854T + RT \ln (1-x)x^3 + (20 \cdot 8 - 0 \cdot 015T)$$
$$\times [x^2 + 3(1-x)^2]$$

for [Q]

$$\Delta\mu = 0 = 160\cdot2 - 0\cdot3029T - 5\cdot50(4) + (0\cdot6 + 0\cdot0832T)$$
$$+(-27\cdot2 + 0\cdot1004T) + RT \ln K$$
$$= 111\cdot6 - 0\cdot1193T + RT \ln x(1-x) + (20\cdot8 - 0\cdot015T)$$
$$\times[x^2 + (1-x)^2]$$

Each of these equilibrium relations defines a line on a $T-x_{CO_2}$ diagram for 4 kbar (figure 6.8(b)). The position of each line is found most easily by substituting a value of x and solving for T. Thus [DI] at $x = 0\cdot5$,

$$0 = 450\cdot0 - 0\cdot5451T - 0\cdot0346T + (20\cdot8 - 0\cdot015T)1\cdot5$$
$$= 481\cdot2 - 0\cdot6022T$$

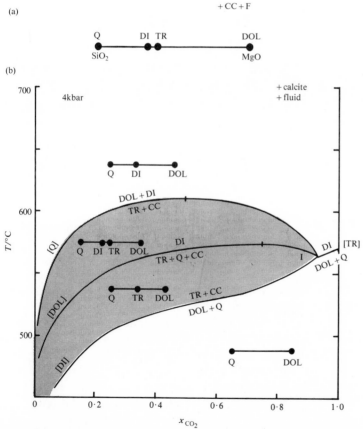

6.8   The compatibility diagram for excess-calcite limestones in the system CaO–MgO–SiO$_2$–H$_2$O–CO$_2$ formed in the presence of a H$_2$O–CO$_2$ fluid phase is shown in (a). The $T-x_{CO_2}$ diagram, (b), shows the phase relations involving TR, DI, DOL and Q at 4 kbar. (WE 6(h))

so the temperature is 799 K or 525 °C. The reactions intersect at the point I. The stable and metastable parts of the reaction lines are assigned using the principle that the metastable part of [i] lies between i-producing reactions (Schreinemaker's rule). The compatibility diagrams are inserted in each T–x_{CO_2} field.

The TR + CC assemblage is restricted to the shaded area in figure 3.20(b). The area is reduced in practice by a reaction involving talc (TA), TA + CC + Q = TR, which intersects [DI] at low temperature; and by a reaction involving forsterite (FO), TR + CC = FO + DI, which intersects [Q] at high temperature. The size of the shaded area depends on the compositions of the phases, the diagram applying only to the phases consisting of their appropriate pure end-member. If iron is added to the system, more Fe will probably enter the TR than the other phases. This will have the effect of reducing the activity of $Ca_2Mg_5Si_8O_{22}(OH)_2$ in the tremolite, moving [Q] to higher temperatures and [DI] to lower temperatures, thus expanding the shaded area. Diluting the fluid phase with another end-member, for example methane, would have the effect of bringing all the reactions down temperature, the amount depending on the amount of dilution of the H_2O and CO_2 in the fluid phase.

Figure 6.8 is used in worked example 8(a) to illustrate some of the processes involved in the metamorphism of limestones.

6(i) Consider the assemblages developed in the model system MgO–SiO_2–CO_2–H_2O at 4 kbar with particular reference to serpentinization and the enstatite–magnesite assemblage.

The phases of interest and for which there are data in appendix A are enstatite (EN), forsterite (FO), talc (TA), magnesite (MAG), quartz (Q) and chrysotile (CH). There are a variety of serpentine minerals, lizardite, antigorite, chrysotile and so on, having similar, but not identical ranges of composition. Although the phase relations among these minerals are no doubt complicated, chrysotile (CH) can be used to represent all the serpentine minerals in calculating the relations between serpentine minerals and other minerals without appreciable error because the Gibbs energy differences between the serpentine minerals (of a particular composition) will be small. Brucite (BR) is omitted initially, but will be introduced later. The only important mineral that has to be omitted because of lack of adequate thermodynamic data is anthophyllite. CO_2-rich parts of the diagram which is going to be constructed might be metastable with respect to anthophyllite-bearing assemblages.

Here, therefore, we have six phases in a system which requires three phases (plus end-members of the fluid phase) to write a balanced reaction. Therefore, three of the six phases must be omitted for each reaction. Each intersection involves four phases, two of the six being omitted. Each intersection can be labelled with these omitted phases in round brackets.

The number of reactions is the number of ways of taking three things from six, which is 20. The number of intersections is 15. In a system with more than one intersection not only will some reactions be metastable, so will some intersections. No stable reactions are involved in a metastable intersection. Although it would be straightforward to plot the positions of the 20 reactions on a T–x diagram, it would be complicated to sort out which intersections are stable and which are metastable. We can take some short cuts.

A good starting point is to assume that Q+MAG is the low temperature assemblage, and that increasing temperature sees the appearance of TA, CH and so on. As FO and EN might be expected to appear at higher temperature than CH and TA, a suitable intersection to start with is (FO, EN). The reactions here are:

[MAG] $CH + 2Q = TA + H_2O$

[TA] $CH + 3CO_2 = 2Q + 3MAG + 2H_2O$

[Q] $2CH + 3CO_2 = TA + 3MAG + 3CO_2$

[CH] $TA + 3CO_2 = 4Q + 3MAG + H_2O$

The qualitative arrangement of the lines around (FO, EN) can be constructed using the shapes of reaction lines on T–x diagrams (figure 3.9) and Schreinemaker's rule (figure 6.9). The compatibility diagrams are MgO–SiO_2 bar diagrams. Using the thermodynamic data in appendix A:

[MAG] $\Delta\mu = 0 = 41\cdot9 - 0\cdot1456T - 1\cdot63P + F_{H_2O} + RT \ln a_{H_2O}$

This reaction cannot take place at a higher temperature than when $x_{H_2O} = 1$, so this fixes the maximum temperature of (FO, EN). [MAG] gives a very low temperature at $x_{H_2O} = 1$, of the order of 0 °C. Nevertheless [Q] and [CH] both extend to higher temperatures and are plotted on figure 6.9. The low temperature of (FO, EN) constrains the positions of some other intersections. For example, (FO, MAG) and (EN, MAG) must both lie on (FO, EN) [MAG], and so both occur at very low temperature. Similarly, (FO, TA) and (EN, TA) must both lie on (FO, EN)[TA] so are either stable at very low temperature (below (FO, EN)) or are metastable at higher temperature. A further constraint is applied by the reaction $EN = FO + Q$ occuring at very high temperatures (of the order of 2000 °C), so that (CH, TA), (CH, MAG) and (TA, MAG) also occur at very high temperatures because they must lie on this reaction. Thus the number of intersections that need to be calculated is reduced.

The stability of the low temperature Q+MAG assemblage is terminated at higher temperature by either $Q + MAG = EN + CO_2$ at (FO, CH) or $Q + 2MAG = FO + CO_2$ at (EN, CH), for which respectively $\Delta G° = 83\cdot8 - 0\cdot1745T - 1\cdot93P + F_{CO_2}$ and $\Delta G° = 173\cdot6 - 0\cdot3498T - 1\cdot38P + 2F_{CO_2}$. At $x_{CO_2} = 1$, the enstatite reaction occurs at 565 °C while the forsterite reaction occurs at 650 °C. This means that the forsterite reaction is metastable with respect to the enstatite reaction, which can now be plotted on figure 6.10. This also means that (FO, CH) is stable while (EN, CH) is metastable. The two other reactions which emanate from (FO, CH) are:

$$TA + MAG = 4EN + CO_2 + H_2O \qquad \Delta G° = 194·2 - 0·3474T - 3·85P$$
$$+ F_{H_2O} + F_{CO_2}$$

$$TA = Q + 3EN + H_2O \qquad \Delta G° = 110·4 - 0·1729T - 1·92P$$
$$+ F_{H_2O}$$

These can be plotted on figure 6.10.

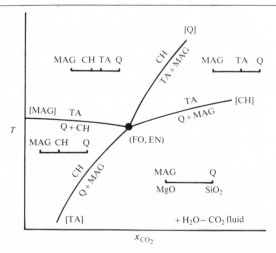

6.9 A qualitative T–x_{CO_2} diagram showing the arrangement of the reaction lines around (FO, EN). (WE 6(i))

The other reaction from (FO, EN) emanating to higher temperature, (FO, EN)[Q], will terminate at either (FO, Q) or (EN, Q), the other being metastable. By the same procedure as for (FO, CH) and (EN, CH), (Q, EN) is stable and (Q, FO) metastable. The three new reactions emanating from (Q, EN) are:

$$5CH = 6FO + TA + 9H_2O \qquad \Delta G° = 687·2 - 1·4238T - 13·68P$$
$$+ 9F_{H_2O}$$

$$TA + 5MAG = 4FO + 5CO_2 + H_2O \qquad \Delta G° = 553·4 - 1·0482T - 10·12P$$
$$+ F_{H_2O} + 5F_{CO_2}$$

$$CH + MAG = 2FO + 2H_2O + CO_2 \qquad \Delta G° = 248·1 - 0·4944T - 4·76P$$
$$+ 2F_{H_2O} + F_{CO_2}$$

The detail of (Q, EN) is shown in the inset on figure 6.10 and is considered later. The second reaction intersects $TA + MAG = EN$ to form the final stable intersection in our temperature range, (Q, CH). The new reactions are:

$$EN + MAG = FO + CO_2 \qquad \Delta G^\circ = 89 \cdot 8 - 0 \cdot 1752T - 1 \cdot 57P + F_{CO_2}$$

$$FO + TA = 5EN + H_2O \qquad \Delta G^\circ = 104 \cdot 4 - 0 \cdot 1722T - 2 \cdot 28P + F_{H_2O}$$

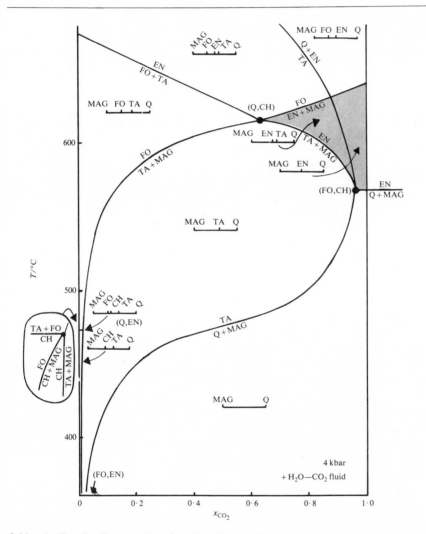

6.10   A T–x_{CO_2} diagram showing the phase relations involving Q, MAG, EN, FO, CH and TA in the system MgO–SiO_2–H_2O–CO_2 in the presence of a H_2O–CO_2 fluid phase. The shaded area gives the field of stability of $EN + MAG$. (WE 6(i))

These are also plotted on figure 6.10. The MgO–SiO_2 compatibility diagrams can then be inserted in each field.

Figure 6.10 has been drawn for each solid phase consisting of the appropriate pure end-member. If one phase is diluted then all the reaction lines which involve that phase will move. Applying such a diagram to a natural assemblage will require all the lines to move by varying amounts. One possibility is that the disposition of the lines will be changed so much that several of the intersections which were stable become metastable, and some which were metastable become stable. This would happen if, for example, we considered the SiO_2 in all the reactions to be in the fluid phase, so that the activity of silica would be much reduced. This would only affect reactions involving Q. One effect could be that (FO, CH) becomes metastable by occurring at a higher temperature than (Q, CH), lying on the metastable extension of $EN = TA + MAG$. This would have the effect of (FO, EN) becoming metastable at a higher temperature than (Q, EN). Certain of the previously metastable intersections would become stable, and so on.

We can make some interesting observations about the assemblages that can form in rocks which are not too far removed from the model system, MgO–SiO_2–H_2O–CO_2, for example the ultramafic rocks, dunites (olivine rocks) and harzburgites (olivine–orthopyroxene rocks). The first is that the stability of CH is restricted to temperatures below 470 °C (at 4 kbar) for very H_2O-rich fluids. The low temperature metamorphism of dunites and harzburgites (for $P_{fluid} = P_{total}$) will only produce serpentine-bearing assemblages for very H_2O-rich fluids. For a small proportion of CO_2 present, the assemblage is $TA + MAG$, while for more CO_2 (but with x_{CO_2} still less than 0·1 below 420 °C) the assemblage is $Q + MAG$. The stability of $EN + MAG$ is restricted to a field around 600 °C for CO_2-rich fluids. It is therefore necessary to change the composition of the metamorphic fluid very substantially in order to form $EN + MAG$ in serpentinites, as well as increase the temperature substantially.

The restrictive field for serpentinization is emphasized by expanding the H_2O edge of the T–x diagram in order to examine the assemblages in very H_2O-rich fluids (figure 6.11). Here two reactions involving brucite (BR) are included because the common serpentinite assemblage is $CH \pm MAG \pm BR$ in terms of the system here. For:

$$BR + CO_2 = MAG + H_2O \qquad \Delta G° = -33·4 + 0·0223T + 0·34P$$
$$+ F_{H_2O} - F_{CO_2}$$

$$CH + BR = 2FO + 3H_2O \qquad \Delta G° = 214·8 - 0·4722T - 4·5P + 3F_{H_2O}$$

There are three shaded fields in figure 6.11. The heavy shading is for the serpentinite assemblage $BR + CH$, the lighter shading for $MAG + CH$, and the light shading for OL+CH. Thus FO = CH + BR and FO = CH + MAG mark the disappearance of olivine in the serpentinization process during cooling, or the re-appearance of olivine in the prograde metamorphism of serpentinites. Thus, serpentinites involving brucite form at temperatures lower than 400 °C (at 4 kbar) for extremely H₂O-rich

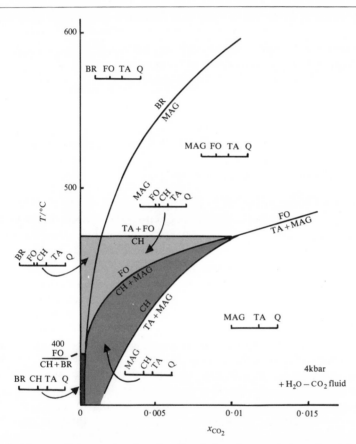

6.11 A T-x_{CO_2} diagram showing details of the phase relations involving BR, MAG, Q, FO, TA and CH in the presence of a very H₂O-rich fluid. Stability range of FO+CH given by light shading; stability of CH+MAG by medium shading; stability of CH+BR by heavy shading. (WE 6(i))

fluids. This temperature decreases little with decreasing pressure until below 0·5 kbar, when the temperature decreases progressively to 180 °C at the Earth's surface. Brucite-free serpentinites can form at temperatures up to 470 °C (at 4 kbar) for slightly less H₂O-rich fluids.

6(j) Use the data tabulated below to find the equilibrium relation for the balanced reaction:

$$FeTiO_3 + Fe_3O_4 = Fe_2O_3 + Fe_2TiO_4$$

ilmenite   magnetite   ilmenite   magnetite

and re-arrange the resulting equation to give a geothermometer.

T/K	1273	1173	1073	973	873
$x_{Fe_2TiO_4,MT}$	0·495 0·617	0·440 0·544	0·340 0·413	0·217 0·255	0·110 0·134
$x_{FeTiO_3,IL}$	0·871 0·908	0·895 0·922	0·914 0·931	0·927 0·940	0·935 0·945

(data from Buddington and Lindsley (1964) for the Fe–Ti–O system at 2 kbar)

At what temperature did an ilmenite–magnetite-bearing lava crystallize if the compositions of the coexisting Fe–Ti oxides are:

$$x_{Fe_3O_4,MT} = 0·5, \qquad x_{Fe_2TiO_4,MT} = 0·45$$

$$x_{Fe_2O_3,IL} = 0·10, \qquad x_{FeTiO_3,IL} = 0·85?$$

Setting up geothermometers/geobarometers for systems for which the available thermodynamic data are insufficient involves fitting experimental data with the equilibrium relation for some appropriate reaction. The equilibrium relation for the above reaction can be set up as a geothermometer which is most useful in calculating temperatures for lavas, coexisting Fe–Ti oxides being common in the groundmass of many basalts and their derivatives (with the exception of alkali basalts and their derivatives).

The two pairs of compositions at each temperature in the above data are for coexisting phases at different oxygen activities. The compositions are for the centres of experimental brackets, with uncertainties of about 0·01 on $x_{FeTiO_3,IL}$ and 0·03 on $x_{Fe_2TiO_4,MT}$. Assuming that both magnetite and ilmenite solid solutions are ideal molecular mixtures then the equilibrium relation for the above reaction in the Fe–Ti–O system at 2 kbar can be written as:

$$0 = a' + b'T + RT \ln\left(\frac{x}{1-x}\frac{1-y}{y}\right)$$

where $x = x_{Fe_2TiO_4,MT}$, $y = x_{FeTiO_3,IL}$. Dividing through by RT and rearranging:

$$\ln K = \frac{a}{T} + b \qquad \text{where} \quad K = \frac{x}{1-x}\frac{1-y}{y}$$

The constants a and b can be calculated from a plot of $\ln K$ against $1/T$.

The uncertainties on $\ln K$ can be calculated from:

$$\sigma^2_{\ln K} = \left(\frac{\partial \ln K}{\partial x}\right)^2_y \sigma^2_x + \left(\frac{\partial \ln K}{\partial y}\right)^2_x \sigma^2_y$$

$$= \left(\frac{0.03}{x(1-x)}\right)^2 + \left(\frac{0.01}{y(1-y)}\right)^2$$

Thus:

T/K	1273	1173	1073	973	873
$\dfrac{1}{T} \times 10^{-4}$	7·86	8·53	9·32	10·28	11·45
$\ln K$	−1·93 −1·81	−2·38 −2·29	−3·03 −2·29	−3·83 −3·83	−4·76 −4·71
$\sigma_{\ln K}$	0·15 0·17	0·16 0·18	0·18 0·20	0·23 0·24	0·35 0·32

These are plotted on figure 6.12. Putting a straight line through these brackets by eye gives:

$$\ln K = -\frac{8200}{T} + 4.6$$

This can be re-arranged to give a geothermometer at 2 kbar:

$$T = \frac{8200}{4.6 - \ln K} \quad (T \text{ in K})$$

The calculation of the uncertainty in a temperature calculated using this equation is set as problem 6(a). A reasonable uncertainty turns out to be about ±30 K. This excludes uncertainties in the formulation of the thermodynamics and any systematic errors in the experimental work.

The crystallization temperature of the ilmenite and magnetite in the lava is 920 °C by substitution into the above equation. If the ilmenite and magnetite are groundmass phases, this is a groundmass crystallization temperature. Even if the Fe–Ti oxides are phenocrysts, the temperature could still be a groundmass temperature if the ilmenite and magnetite continued to equilibrate down to that temperature, as seems likely from the results of the application of this geothermometer to many igneous coexisting Fe–Ti oxides.

6(k) Calculate the temperature of formation of a magnetite(MT)- and quartz(Q)-bearing metamorphic rock from the following oxygen isotope information:

$$\delta^{18}O_Q - \delta^{18}O_{MT} = 9 \quad \text{where} \quad \delta^{18}O_i = \left(\frac{\left(\dfrac{^{18}O}{^{16}O}\right)_i}{\left(\dfrac{^{18}O}{^{16}O}\right)_{SMOW}} - 1\right) \times 10^3$$

given that the $\Delta G°$ for the reaction:

$$\tfrac{1}{2}Si^{16}O_2 + \tfrac{1}{4}Fe_3{}^{18}O_4 = \tfrac{1}{2}Si^{18}O_2 + \tfrac{1}{4}Fe_3{}^{16}O_4$$

is:

$$\Delta G° = -\frac{46\cdot 31}{T} \qquad (T \text{ in K})$$

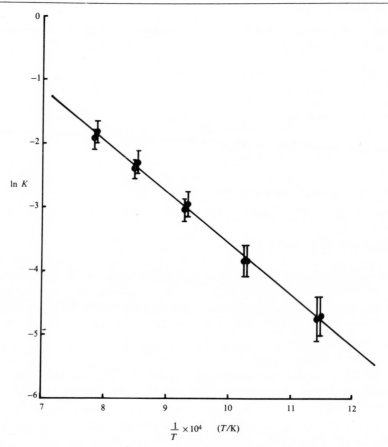

6.12 A $\ln K - 1/T$ diagram used to calibrate the Fe–Ti oxide geothermometer in WE 6(j).

Oxygen isotope geothermometry, of which this is an example, involves the application of equilibrium thermodynamics to the distribution of the oxygen isotopes ^{18}O and ^{16}O between phases. The fractionation of these isotopes between each mineral pair is a function of temperature. After experimental calibration, each can be set up as a geothermometer.

The δ notation is used because the fractionations involved are very small, and this notation exaggerates the small difference in isotope ratio between the $^{18}O/^{16}O$ for the mineral and the standard, SMOW, standard mean ocean water. The equilibrium relation for the above isotopic exchange reaction is:

$$0 = \Delta G° + RT \ln \left(\frac{^{18}O}{^{16}O}\right)_Q \Big/ \left(\frac{^{18}O}{^{16}O}\right)_{MT}$$

$$= \Delta G° + RT \ln \alpha$$

where α, the fractionation factor, is identically equal to the equilibrium constant (at equilibrium). Now, re-arranging the definition of α:

$$\left(\frac{^{18}O}{^{16}O}\right)_i = \left(\frac{^{18}O}{^{16}O}\right)_{SMOW} (\delta^{18}O_i \times 10^{-3} + 1)$$

Also, α is usually very close to 1, so that $\ln \alpha \simeq \alpha - 1$. Combining:

$$\ln \alpha \simeq \alpha - 1 = \frac{\delta^{18}O_Q - \delta^{18}O_{MT}}{\delta^{18}O_{MT} + 1000} \simeq \frac{\delta^{18}O_Q - \delta^{18}O_{MT}}{1000}$$

as $\delta^{18}O_{MT}$ is small compared with 1000. Thus:

$$1000 \ln \alpha = \delta^{18}O_Q - \delta^{18}O_{MT} \equiv \Delta_{Q\text{-}MT}$$

These must also equal:

$$-\frac{1000 \Delta G°}{RT}$$

Oxygen isotope geothermometers are usually quoted in the form:

$$1000 \ln \alpha_{XY} = A + \frac{B}{T^2}$$

for the fractionation between phases X and Y. The required analytical measurements are the isotope ratios, so that the temperature can be calculated using:

$$\Delta_{X\text{-}Y} = \delta^{18}O_X - \delta^{18}O_Y = A + \frac{B}{T^2}$$

or, by re-arranging:

$$T = \sqrt{\frac{B}{\Delta_{X\text{-}Y} - A}}$$

In this case:

$$1000 \ln \alpha = -\frac{1000}{RT}\left(-\frac{46\cdot31}{T}\right) = \frac{5\cdot57 \times 10^6}{T^2}$$

Substituting $\Delta_{Q-MT} = \delta^{18}O_Q - \delta^{18}O_{MT} = 9$ (or $\alpha = 1{\cdot}00904$) gives $T = 515\,°C$. A temperature determined by oxygen isotope geothermometry will refer to the temperature at which isotopic equilibrium was frozen in, which will depend on the cooling rate, grain size and so on (presuming equilibrium was actually attained at some time during the history of the assemblage). The uncertainty on the temperature due to the uncertainty in Δ (assuming that A and B are reasonably well known) can be calculated using:

$$\sigma_T = \left| \frac{dT}{d\Delta_{Q-MT}} \right| \sigma_{\Delta_{Q-MT}} = \frac{T^3}{2B}\, \sigma_{\Delta_{Q-MT}}$$

Assuming $\sigma_{\Delta_{Q-MT}} = 0{\cdot}3$, this gives an uncertainty on the temperature of $13\,°C$. Obviously the temperature uncertainty will be larger if the value of B for the geothermometer is smaller (for a particular value of σ_Δ). The Q–MT geothermometer has the largest value of B of all the geothermometers which have been calibrated so far.

Problems 6

6(a)   A geothermometer has the form:

$$\ln K = -\frac{a}{T} + b \quad \text{where} \quad K = \frac{x(1-y)}{(1-x)y}$$

where x and y are composition terms and $a = 8200 \pm 100$ and $b = 4{\cdot}6 \pm 0{\cdot}1$. What is the uncertainty on the calculated temperature if:

  (a) $x = 0{\cdot}4$ and $y = 0{\cdot}9$;
  (b) $x = 0{\cdot}4 \pm 0{\cdot}01$ and $y = 0{\cdot}9 \pm 0{\cdot}01$?

6(b)   Given that a $0{\cdot}5$ wt% uncertainty in the SiO_2 for the amphibole in WE 4(b) contributes most to the uncertainties in the recalculated amounts of the other cations, is there any Na on M4?

6(c)   A metamorphic limestone contains the assemblage quartz (Q), talc (TA), calcite (CC), dolomite (DOL), sphene (SPH) and rutile (RUT). Calculate the T and x_{CO_2} for this assemblage at 2 kbar using the intersection of the curves for the reactions:

  (1) $TiO_2 + CaCO_3 + SiO_2 = CaTiSiO_5 + CO_2$
        RUT   CC   Q   SPH   F
  (2) $3CaMg(CO_3)_2 + 4SiO_2 + H_2O = Mg_3Si_4O_{10}(OH)_2 + 3CaCO_3 + 3CO_2$
        DOL   Q   F   TA   CC   F

For reaction (1), assume that $\Delta G°$ is known within $\pm 1\,kJ$, whereas for reaction (2), assume $\Delta G°$ is known within $\pm 3\,kJ$. Use unit activities for SiO_2 in Q and TiO_2 in RUT, and:

$$a_{Mg_3Si_4O_{10}(OH)_2,TA} = 0.75$$

$$a_{CaMg(CO_3)_2,DOL} = 0.95$$

$$a_{CaCO_3,CC} = 0.95$$

$$a_{CaTiSiO_5,SPH} = 0.95$$

Assume that the metamorphic fluid contained only H_2O and CO_2.

References

Wood, B. J. and Fraser, D. G., 1976, *Elementary Thermodynamics for Geologists*, Oxford University Press, London: Chapters 1–4 are a different development to the same end— thermodynamic calculations on mineral assemblages.

Powell, R. and Powell, M., 1977, Geothermometry and oxygen barometry using coexisting iron-titanium oxides: a reappraisal. *Min. Mag.*, **41**, 257–63: Source for WE 6(j).

Javoy, M., 1977, Stable isotopes and geothermometry. *J. Geol. Soc. Lond.*, **133**, 609–36: Good further reading on oxygen isotope geothermometry.

Chapter 7

Processes 1

In the next two chapters, processes are considered which are involved in the generation of equilibrium assemblages (nucleation and growth, closure), and processes which affect the evolution of a rock or series of rocks (buffering, metasomatism, fractionation). These latter processes can all be considered to involve the maintenance of equilibrium throughout the process, but in each case there is a specific restriction on the scale over which equilibrium is attained or the way equilibrium is attained. Many of the considerations are qualitative (with the help of phase diagrams) as the mathematics tends to be difficult, and essentially qualitative due to lack of the appropriate information to use with it.

Nucleation and growth

The nucleation and growth of a mineral in an assemblage will only occur if the formation of that mineral lowers the Gibbs energy of the assemblage. However, even when the new mineral has become stable, it will not necessarily nucleate. This can be seen with the help of the $G-x$ diagram in figure 7.1. Nucleation of a phase comes about as the result of fluctuations in composition and the positions of the atoms in the system, if the presence of this phase in the system lowers the Gibbs energy. The Gibbs energy of a mole of nucleii is larger than the Gibbs energy of a mole of large crystals for a particular composition because there is a contribution to the Gibbs energy from the surfaces of the nucleii. This contribution is given by the surface area of the nucleii times a constant,

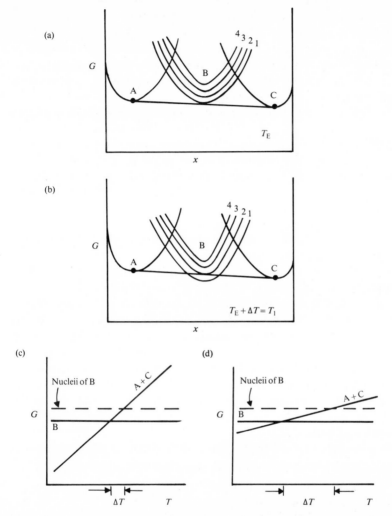

7.1 The $G-x$ diagrams, (a) and (b), show that B will appear in A + C assemblages at T_1 rather than T_E, where the numbers on the $G-x$ curves for B refer to: 1, large crystals; 2, nucleii on grain boundaries; 3, nucleii on dislocations in crystals; and 4, nucleii in 'good' crystal. Nucleii will appear on grain boundaries at T_E. The $G-T$ diagrams, (c) and (d), show that the overstepping of the reaction, ΔT, depends on the angle of divergence of the lines for B and A + C, for a given ΔG for nucleation of B.

the surface energy, S. Therefore nucleation will only occur when the $G-x$ loop for nucleii of the new phase intersects the previously stable common tangent on a $G-x$ diagram. There are usually several locations in a system where nucleation can occur; inside grains (on dislocations or impurities, or in 'good' crystal), on grain boundaries (between minerals A and A, minerals A and B, and so on), on triple junctions (intersections of three minerals), and so on. The $G-x$ loop for each of these will be slightly different, shown schematically in figure 7.1. The dominant site of nucleation will be at the location which gives the lowest Gibbs energy loop. The practical result of the new phase appearing in the system when the $G-x$ loop for the nucleii crosses the common tangent is that the reaction involved has been overstepped, the reaction taking place at T_1 rather than T_E. The amount of this overstepping will depend on how rapidly the $G-x$ loop for the new phase is moving relative to the $G-x$ loops of the other phases with temperature. This depends on the temperature dependence of $\Delta\mu$ for the reaction. Two cases are presented in figures 7.1(c) and (d). In (c), a large temperature dependence, as found in most dehydration and decarbonation reactions, means that the reaction is overstepped by a small amount, ΔT being probably less than $10\,K$ at metamorphic conditions. In (d), a small temperature dependence as found in many solid–solid reactions may result in the substantial overstepping of the reaction. This may be part of the reason for the 'misbehaviour' of aluminosilicates in high-grade meta-pelites. Another aspect of the overstepping of the reaction is that further overstepping may be required to get a substantial nucleation rate. The importance of this will depend on the difference between the rate of change of conditions and the rate of nucleation. The rate of nucleation may be the limiting factor in the development of equilibrium assemblages in sedimentary environments, in rapidly cooled magmas, retrograde metamorphism, and so on. It may also be very important in the experimental determination of reaction equilibria. The overstepping effect may be minimized or even reversed if the minerals have stored Gibbs energy in the form of strain. The presence of this strain energy is the reason why nucleation and growth of unstrained minerals occur in deformed rocks.

Once nucleii are formed, they must be able to grow before we will be able to observe the new mineral in an assemblage. To a large extent, growth depends on the supply of the appropriate elements to the nucleii. It might turn out that although most nucleii were being formed inside grains, nucleii on **grain** boundaries grew because the supply of materials to the grain boundary nucleii was more effective than the supply to nucleii within grains. Supply may not be sufficient to ensure growth because the surfaces of the growing nucleii may become 'poisoned' resulting in the inhibition of grain growth. Carbon poisoning of the surfaces of grains may occur in graphitic rocks during metamorphism, resulting in the smaller grain size and preservation of strained grains in graphitic rocks.

Growth occurs in the continued attempt of the system to lower its Gibbs energy, a small contribution to the Gibbs energy of the system coming from each grain boundary. The positions of the grain boundaries tend to migrate to produce equilibrium arrangements, i.e. those minimizing the

Gibbs energy of the system. These arrangements depend on the surface energy properties of the minerals, which are different for different directions in the crystal. Each mineral has a preferred shape as indicated by mineral habit in vugs. A balance is obtained between these preferred shapes when minerals crystallizing together impinge on each other. The Dupré equation relates the equilibrium angles at the intersection of grain boundaries (figure 7.2):

$$\frac{S_{23}}{\sin \theta_1} = \frac{S_{13}}{\sin \theta_2} = \frac{S_{12}}{\sin \theta_3} \tag{7.1}$$

where S_{ij} is the surface energy for a boundary between minerals i and j, effectively the Gibbs energy of a unit area of grain boundary between i and j. The surface energy will depend also on the crystallographic orientation of the grains as no mineral has isotropic surface properties.

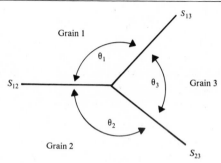

7.2   A section through grain boundaries between phases 1, 2 and 3 showing the labelling of angles, θ_i, and the surface energies, S_{ij}.

This is one reason why most grain boundaries are curved. However, (7.1) can be used in a simple-minded way to show features of grain shapes assuming isotropic surface properties. If the junction is between three grains of the same mineral (or different minerals with similar surface energies) then $S_{12} = S_{23} = S_{13}$. This means that $\theta_1 = \theta_2 = \theta_3$, and as $\theta_1 + \theta_2 + \theta_3 = 360°$, then $\theta_1 = \theta_2 = \theta_3 = 120°$. Departures from the 120° angles are caused by the differences in the surface energies for the different mineral pairs. Equilibrium grain boundary configurations are the familiar triple junctions found particularly in high-grade metamorphic rocks like granulites. Their presence indicates that the system had plenty of time to reach equilibrium.

The Gibbs energy of a mineral in a stress field depends on the orientation of the crystal lattice to the principle stresses. Therefore another aspect of the minimization of Gibbs energy of a system of grains concerns the orientation of the crystal lattices of the grains. This is one way of accounting for the preferred orientation of minerals in metamorphic rocks which crystallized in a stress field.

Mass transfer

In the equilibrium thermodynamics chapter we saw that equilibrium between two phases A and B involved, for example, $\mu_{1A} = \mu_{1B}$. Now if we consider adjacent subsystems A and B in a rock which have μ_{1A} different from μ_{1B}, then the system will try and come to equilibrium by removing the chemical potential gradient between A and B. If the two subsystems contain a fluid phase, then the gradient can be equalized by the mass movement of the fluid (infiltration) due to a temperature and/or pressure gradient between the subsystems, accompanied by reaction between the fluid and the solid phases. If the fluid is stationary or if there is no fluid phase, then the chemical potential gradient can be equalized by the movement of atoms (species, molecules) between the two subsystems, again accompanied by reaction between these atoms and the solid phases. Movement as a response to a chemical potential gradient is known as diffusion. At least below about 800 °C, this movement is much larger along grain boundaries (grain boundary diffusion), particularly in the presence of a fluid phase, than through grains (volume diffusion). The rate of movement varies from species to species and is a function of temperature in each of the possible environments of movement. Diffusion and infiltration are mass transfer processes because the compositions of the subsystems inevitably change during their operation.

Most rocks are compositionally heterogeneous on one scale or another. This heterogeneity is often reflected in chemical potential gradients between the different bands or parts of the rock. Diffusion is and has been continually trying to remove these chemical potential gradients. Note that the removal of a chemical potential gradient does not imply making the compositions of the bands the same. A result of the removal of all chemical potential gradients between compositionally banded rocks at a particular temperature and pressure would be that each band would contain the same mineralogy, with each of the minerals in every band having the same composition. Nevertheless the proportions of the minerals in each of the bands would be different, so the compositions of the bands would be different.

It is clear that mass transfer was not sufficiently rapid to remove chemical potential gradients over a distance of millimetres or centimetres even over the time span of a regional metamorphic event as metamorphic rocks are usually banded on this scale, with different mineralogies and mineral chemistry in adjacent bands. At much lower pressures where interconnecting pores are present through which fluids can flow, infiltration is likely to be the most effective mass transfer process, and could operate over much larger distances. Diffusion seems to be most important on the scale of a few millimetres to centimetres enabling mineral growth and probably controlling grain size in diagenesis, metamorphism and igneous crystallization.

Conspicuous evidence of mass transfer on a large scale, usually in the form of reaction zones between dissimilar rock units, occurs in few geological environments, the best known being the development of skarns between limestone and granite. Conspicuous evidence on a small scale comes where two minerals which were no longer stable together react forming reaction zones (coronas, kelyphitic rims) between them. This is exactly equivalent to dissimilar rock units forming reaction zones between them. Note that the above evidence for mass transfer will only be conspicuous if the process is halted before completion (i.e. before one rock unit or one mineral is destroyed).

Diffusion can be considered in terms of mobility, say in terms of the distance an element can move in unit time under the action of a unit chemical potential gradient for the conditions of interest (pressure, temperature, grain size, fluid composition, and so on). Some elements will be much more mobile than others. Elements which can move over distances much larger than the size of the system for the conditions of interest are sometimes called 'perfectly mobile', others which can move only very small distances within the system are called 'inert'. Alternatively, the system can be considered to be 'open' to the former elements, while the system is 'closed' to the latter elements. Note that an element which is perfectly mobile in one situation may be inert in another situation.

The way the mobility of an element affects the mineralogy of a rock can be seen by considering figure 7.3. Most compositions in the binary system at this P–T will contain two phases, while most μ_{MgO} values will stabilize only one phase. Thus if MgO is mobile and its chemical potential controlled externally from the system (i.e. superimposed on the system), the system is most likely to contain only one phase rather than the customary two. (This can be generalized for systems with n end-members. For most compositions there are n phases, but if m independent end-members are mobile and have their chemical potentials superimposed on the system then there are usually $n-m$ phases.) Note that in a binary system, the behaviour of the two chemical potentials is sympathetic which means that fixing the chemical potential of MgO fixes the chemical potential of SiO_2, without implying that SiO_2 is necessarily mobile.

Consider two bands of rock abutting each other, one containing just forsterite (μ_{MgO} at A), the other containing just quartz ($\mu_{MgO} = 0$, at B), (figure 7.3). This corresponds to dunite against quartzite or quartz against forsterite. If MgO is much more mobile than SiO_2, then MgO will diffuse across the contact and convert quartz to enstatite (figure 7.4), assuming that the rate of attainment of equilibrium is more rapid than the diffusion rate. The velocity of encroachment of the enstatite zone on the quartz will depend on the efficiency of the diffusion process, how easily MgO can diffuse through the growing enstatite zone. The monomineralic nature of the enstatite zone is a direct consequence of figure 7.3(b), the chemical potential gradient from the quartz to the forsterite only allows enstatite to occur adjacent to quartz or forsterite at the appropriate zone boundary. The chemical potential gradient across the reaction zone is the driving force for the continuing growth of enstatite. In the case of figure 7.4(b), SiO_2 is much more mobile than MgO, so the enstatite zone moves into the forsterite. In the case of figure 7.4(c), both MgO and SiO_2 are mobile, the MgO being more mobile than the SiO_2, so that the enstatite zone

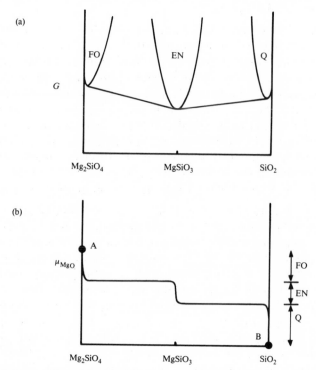

7.3 A schematic G–x diagram, (a), and corresponding μ_{MgO}–x diagram, (b), for the system MgO–SiO_2 involving FO, EN, and Q.

encroaches faster on the quartz than on the forsterite. Obviously to predict the disposition of the enstatite zone and its rate of growth, we would need to use the mathematics of diffusion. This can be very complicated so it will not be attempted here; sufficient is achieved for our purposes by considering diffusion in a qualitative way.

Whereas the zone sequence developed in a binary system is defined at a particular temperature and pressure by the stability of the phases, the situation gets progressively more complicated as further end-members are added. Some of the possibilities can be seen by considering the reaction between a wet quartzite A and a dry peridotite B in the ternary system MgO–SiO_2–H_2O (figure 7.5).

The zone sequence is well defined if only one of the end-members is mobile (or one of the end-members is much more mobile than the

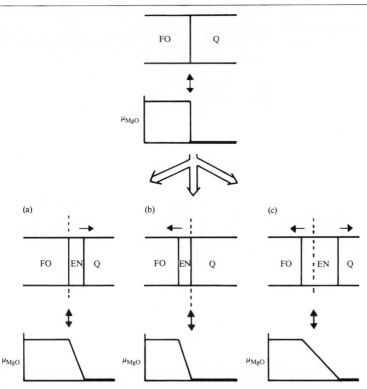

7.4 Diffusion-generated EN reaction zone between FO and Q assuming: (a) MgO is mobile; (b) SiO_2 is mobile; and (c) both MgO and SiO_2 are mobile. The corresponding chemical potential gradients are shown in each case.

others). In this case the ratio of the amounts of the other two end-members must remain constant in the two rocks throughout the diffusion process. If, in figure 7.5, H_2O is the mobile end-member, whilst SiO_2 and MgO are relatively immobile, then the composition of B will move towards the H_2O apex along the broken line. The movement along this broken line will be in jumps because only two phases can occur in each of the reaction zones, three being present only across zone boundaries (figure 7.5(b)). The sequence of zones is peridotite, enstatite + chrysotile, talc + chrysotile, then quartzite + H_2O. These zones will all be formed in the peridotite because the quartzite is supplying the H_2O to the peridotite. Note that the EN + CH and CH + TA zones do not have a fluid phase even though H_2O is diffusing through them. If the peridotite occurs as a

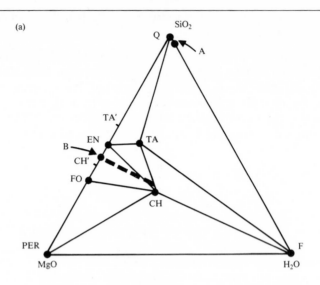

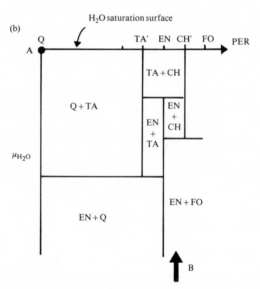

7.5 A MgO–SiO$_2$–H$_2$O compatibility diagram, (a), with the quartzite, A, and the peridotite, B, marked; and the path (broken line) showing the direction the composition of B changes as the result of H$_2$O diffusion from A. (b) represents the sequence of reaction zones developed for a particular μ_{H_2O} gradient for each MgO/SiO$_2$ composition. The arrow marks the composition of the peridotite. A vertical line from this arrow to the H$_2$O saturation surface gives the reaction zone sequence EN + CH, TA + CH between the peridotite and the quartzite.

small mass in the quartzite, the peridotite will be made over to TA + CH + H_2O if the process is allowed to operate long enough.

If SiO_2 and H_2O are mobile then the zone sequence will depend on the relative mobilities of SiO_2 and H_2O. Possible zone sequences can be seen on a μ–μ diagram (figure 7.6). μ–μ diagrams are equivalent to the ln a–ln a diagrams used previously. Like ln a–ln a diagrams, slopes of reaction lines can be obtained from reaction coefficients. The reaction between chrysotile, enstatite and the mobile end-members SiO_2 and H_2O is:

$$2H_2O + 3MgSiO_3 = Mg_3Si_2O_5(OH)_4 + SiO_2$$

The slope of the line for this reaction on a μ_{SiO_2}–μ_{H_2O} diagram is $-n_{H_2O}/n_{SiO_2} = 2$. A qualitative μ–μ diagram which is sufficient for our purposes can be drawn knowing the compatibility diagram for a certain pressure and temperature for those conditions, using Schreinemaker's rule. The quartz and H_2O saturation lines are also included on figure 7.6, as are the positions of the quartzite, A, and the peridotite, B. Comparing figure 7.6 and figure 7.5(a) we can see the differences between composition and μ–μ diagrams. In figure 7.5(a), most compositions contain three phases, in figure 7.6 most pairs of μ values involve only one phase. Thus assemblages should be monomineralic for superimposed chemical potentials (and chemical potential gradients).

The two extremes of H_2O and SiO_2 mobility are H_2O much more mobile than SiO_2 and vice versa. The former of these, already considered above, is given by the path abcd, the latter by the path ae, the only zone developed being an enstatite monomineralic zone (corresponding to the binary case considered above). If SiO_2 and H_2O are equally mobile then two monomineralic zones involving enstatite and talc are developed, the chemical potential gradients across the zones being given by af. In this case, if the process goes to completion, a small peridotite body would be completely replaced by talc + quartz + fluid. Zone sequences intermediate between these two extremes will be developed for different relative mobilities of H_2O and SiO_2.

Zone sequences can be visualized on μ–μ diagrams as long as not more than two (preferably one) end-members are mobile. The sequences in more complicated situations can be predicted using the mathematics of diffusion if information is available on the mobilities of the end-members, and so on.

For diffusion to be an effective process, not only has a chemical potential gradient to be present and the mobilities of the end-members sufficient, but the amount of the end-member in the medium through which the diffusion is taking place must be large enough so that the mass transfer is sufficiently rapid. This is particularly important for diffusion in a grain boundary fluid phase. If the solubility of an end-member in the fluid is very small, then the amount of that end-member which can be moved in a particular time will be small, regardless of the mobility of that end-member. The low solubility of most end-members in the fluid phase is probably one of the reasons why diffusion structures on a hand-specimen or outcrop scale are not observed more frequently in metamorphic rocks. The most mobile end-members should be H_2O and CO_2 so that one might expect any chemical potential gradients for these end-members to have been erased on an outcrop scale during metamorphism; but this is certainly not always the case, most noticeably in metamorphic limestones.

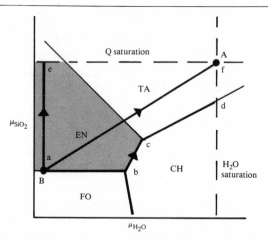

7.6   A qualitative $\mu_{H_2O}-\mu_{SiO_2}$ diagram showing different diffusion paths between the quartzite, A, and the peridotite, B, depending on the relative mobilities of H_2O and SiO_2.

Solubility, with the amount of fluid flow, also limits the effectiveness of mass transfer by infiltration. Reaction zones formed by infiltration can be similar to reaction zones formed by diffusion, particularly in complicated systems, although the behaviour of solid solutions can be distinctively different.

Temperature–time dependence of rate processes

Rate processes like nucleation, growth and diffusion are usually strong functions of temperature, the rates increasing with temperature. This has important consequences with respect to rocks following a temperature–pressure–time path, particularly igneous and metamorphic rocks.

Consider the equilibrium between orthoclase and albite solid solutions across a grain boundary. The equilibrium compositions of the coexisting phases at any temperature can be seen from figure 7.7. The two feldspars have formed by unmixing at T_1. While the rate of equilibration between the feldspars is fast compared with the cooling rate, the compositions of the feldspars will follow the limbs of the solvus. However the rate of equilibration will decrease with decreasing temperature until equilibrium cannot be attained at a particular temperature, T_a, in the time available. With further cooling the equilibrium is approached less and less until below a certain temperature T_b, the feldspars do not change composition further, their resultant compositions being $x_{Na,Or}$ and $x_{Na,Ab}$. These are

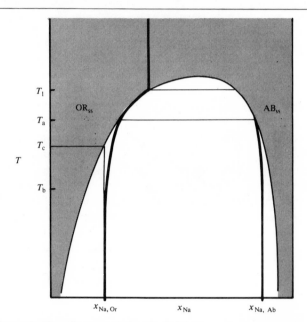

7.7   A schematic T–x diagram of the alkali feldspar solvus illustrating the idea of closure temperatures, T_c; see text.

the compositions which we would measure on an electron microprobe. These compositions could have been in equilibrium at T_c, which can be called a closure temperature. It is a minimum estimate of the temperature at which the feldspars were last in equilibrium (T_a), and a maximum estimate of the temperature at which the feldspars last changed composition (T_b). Closure temperatures are dependent on cooling rates, faster cooling giving higher closure temperatures, slower cooling lower closure temperatures.

The situation is usually more complicated than in the above. As the feldspars have a finite size, equilibrium between the grains involves diffusing Na and K into/out of the orthoclase and albite solid solutions. 'Rate of equilibration' must include this effect. The result is that the interiors of the grains stop equilibrating before the outer parts of the grains. Each part of the grain follows its own T–x path resulting in zoned grains. The amount of this zoning depends on the importance of the rate of Na–K diffusion in the feldspars in the rate of equilibration. The zoning may be too fine to resolve by electron microprobe in which case the closure temperature determined from the analyses of the grains will be some sort of average of the closure temperatures of the different parts of the grain. If these effects are important then the closure temperature from the orthoclase and the albite solid solutions will be different, i.e. the temperature corresponding to the intersection of each feldspar composition with the limbs of the solvus will be different.

In much of geothermometry and geobarometry we are measuring closure conditions, as the mineral assemblages followed some temperature–pressure–time path, with equilibrium usually attained along part of it. The value of a particular geothermometric/geobarometric method will depend on how nearly the temperature/pressure estimates from the method are to the actual conditions which we are trying to discover. For example coexisting sulphide and coexisting Fe–Ti oxide geothermometers give temperatures of the order of 600 °C or less for many basic layered intrusions, reflecting low closure temperatures compared to the igneous crystallization at temperatures greater than 1000 °C. Geothermometry/geobarometry methods which are critically dependent on the assemblage, and less on the compositions of the phases, are less vulnerable in this respect because the development of lower temperature assemblages will involve nucleation and growth of new phases, which seems to take place tardily in most metamorphic rocks, and when it has taken place the new minerals can often be recognized for what they are.

Worked examples 7

7(a) How does a muscovite–quartz–kyanite–H_2O rock evolve on being slowly heated from the kyanite field at (a) 7 kbar, and (b) 4 kbar, assuming that reactions tend to be overstepped by about 1 kJ to allow for nucleation and growth of the products.

The main reactions of interest are the reactions between the aluminosilicate polymorphs and the muscovite + quartz breakdown reaction (figure 6.5). At 7 kbar for the kyanite = sillimanite reaction (WE 6(d)):

$$\Delta\mu = 13\cdot33 - 0\cdot0140\,T \text{ kJ}$$

If the reaction is overstepped by 1 kJ, the temperature of appearance of sillimanite will be $T = (13\cdot33 + 1)/0\cdot014 = 750\,°C$, involving overstepping the equilibrium temperature of the reaction by 70 °C. At 7 kbar for the muscovite + quartz breakdown reaction in the end-member system (WE 6(e)):

$$\Delta\mu = 78\cdot2 - 0\cdot0775\,T$$

The temperature of appearance of the products will be $T = (78\cdot2 + 1)/0\cdot0775 = 745\,°C$, overstepping by $10\,°C$. Thus, on this logic, sillimanite will first appear in this rock associated with the breakdown of muscovite + quartz. This sillimanite will nucleate in the energetically most favourable site, which in pelites does not appear to be in the kyanite. Kyanite will be replaced by sillimanite at a slightly higher temperature.

At 4 kbar, for the kyanite = andalusite reaction:

$$\Delta\mu = 5\cdot95 - 0\cdot0079\,T$$

and for the andalusite = sillimanite reaction:

$$\Delta\mu = 5\cdot65 - 0\cdot0061\,T$$

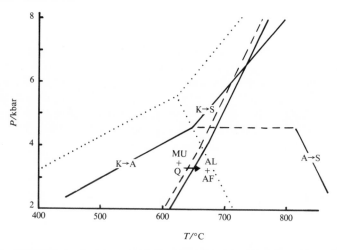

7.8   The effective P–T phase diagram (full lines) compared to the equilibrium phase diagram (dotted line for Al_2SiO_5, broken line for MU + Q breakdown) for heating a quartz–muscovite–kyanite–H_2O assemblage, assuming 1 kJ overstepping of each reaction. (WE 7(a))

The first reaction is overstepped by $125\,°C$, the second by $165\,°C$. At 4 kbar, the quartz + muscovite reaction oversteps by $10\,°C$. Thus on this logic, andalusite will replace kyanite at a temperature of about $600\,°C$. Sillimanite appears first at about $675\,°C$ associated with the breakdown of muscovite + quartz. The andalusite would only be replaced by sillimanite at temperatures of the order of $820\,°C$.

These relations are summarized in figure 7.8. Above $6\cdot5$ kbar, sillimanite will appear in pelitic assemblages by the muscovite + quartz breakdown reaction before the reaction kyanite = sillimanite. Between $4\cdot5$ and $6\cdot5$ kbar sillimanite appears at the expense of kyanite. Textures in high-grade pelites indicate that the reaction kyanite = sillimanite does not occur directly by the replacement of kyanite by sillimanite, but instead by a complex set of subreactions involving biotite, muscovite and quartz, the subreactions summing to the reaction kyanite = sillimanite.

The above discussion is no doubt an oversimplification: the amount of overstepping is likely to depend on temperature and the reaction, the distinction between fibrolitic and crystalline sillimanite is probably significant, the overstepping of the reaction will be less if the kyanite is strained, and so on. Nevertheless the approach does give some insight into the effects of allowing for an energy input to permit nucleation and growth.

7(b) Three minerals, A, B and C, with isotropic surface energies ($S_{AB} =$ 800, $S_{AC} = 500$, $S_{BC} = 400$, $S_{AA} = S_{BB} = S_{CC} = 600$) occur in a rock. Draw the equilibrium grain boundary configurations.

The grains should have straight grain boundaries as the minerals are taken to have isotropic surface energy properties. To calculate the dihedral angles (θ_i) requires some re-arrangement of the Dupré equation (7.1). The starting point is the two independent equations in (7.1), slightly re-arranged:

$$0 = \sin \theta_1 - \frac{S_{23}}{S_{13}} \sin \theta_2$$

$$\sin \theta_3 = \frac{S_{12}}{S_{23}} \sin \theta_1$$

Now, by standard trigonometric manipulations,

$$\begin{aligned}
-\sin \theta_2 &= -\sin (360 - \theta_1 - \theta_3) \\
&= \sin (\theta_1 - \theta_3) \\
&= \sin \theta_1 \cos \theta_3 + \cos \theta_1 \sin \theta_3 \\
&= \sin \theta_1 (1 - \sin^2 \theta_3)^{1/2} + \cos \theta_1 \sin \theta_3
\end{aligned}$$

Substituting this into the first equation, and then substituting the second equation (for $\sin \theta_3$) into the result gives:

$$0 = \sin \theta_1 + \frac{S_{23}}{S_{13}} \left\{ \sin \theta_1 \left[1 - \left(\frac{S_{12}}{S_{23}}\right)^2 \sin^2 \theta_1 \right]^{1/2} + \frac{S_{12}}{S_{23}} \cos \theta_1 \sin \theta_1 \right\}$$

Cancelling $\sin \theta_1$, substituting $\sin^2 \theta_1 = 1 - \cos^2 \theta_1$ and re-arranging:

$$\cos \theta_1 = \frac{1}{2} \left(\frac{S_{23}^2 - S_{12}^2 - S_{13}^2}{S_{12} S_{13}} \right)$$

$$\cos \theta_2 = \frac{1}{2} \left(\frac{S_{13}^2 - S_{12}^2 - S_{23}^2}{S_{12} S_{23}} \right)$$

and:

$$\theta_3 = 360 - \theta_1 - \theta_2$$

Substituting into these equations:

For an ABC junction $\theta_A = 156°$, $\theta_B = 149°$, $\theta_C = 55°$

ABB	$\theta_A = 136°$, $\theta_B = 112°$
BAA	$\theta_B = 136°$, $\theta_A = 112°$
ACC	$\theta_A = 106°$, $\theta_C = 129°$
CAA	$\theta_C = 106°$, $\theta_A = 129°$
BCC	$\theta_B = 83°$, $\theta_C = 139°$
CBB	$\theta_C = 83°$, $\theta_B = 139°$

These can be assembled into part of an imaginary thin section involving these phases (figure 7.9). Grain boundaries inevitably will be slightly curved so that the equilibrium triple junction configurations can be accommodated throughout the rock.

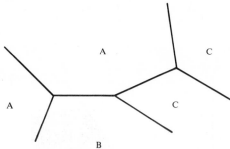

7.9 Part of an imaginary thin section involving phases A, B and C; these equilibrium grain boundary configurations being calculated using the surface energies in WE 7(b).

The effect of AC and BC surface energies being much less than the AB surface energy is to make θ_C small. If the value of S_{AB} is increased relative to $S_{AC} + S_{BC}$, there will be no equilibrium grain boundary configuration involving A, B and C, if S_{AB} is greater than $S_{AC} + S_{BC}$. In this case, C will 'wet' the grain boundaries of A and B. This situation is relevant in considering the behaviour of aqueous fluids during metamorphism and silicate liquids during partial melting. For example, the separation of a partial melt from its source would be easier and could take place at a lower degree of partial melting, if the surface energies of liquid–mineral boundaries were small so that the silicate liquid wetted the mineral boundaries. During metamorphism, the development of triple junctions between minerals would not necessarily be observed if there existed a fluid phase along all the grain boundaries during crystallization because the grain boundary configurations would be controlled by interaction between the minerals and the fluid, not by interaction between the minerals.

7(c) Consider the interaction of a peridotite (enstatite + forsterite) with a quartzite which contains a CO_2 fluid phase if (a) just CO_2 is mobile, (b) both CO_2 and SiO_2 are mobile. The compatibility diagram for the conditions of interest is given by figure 7.10(a).

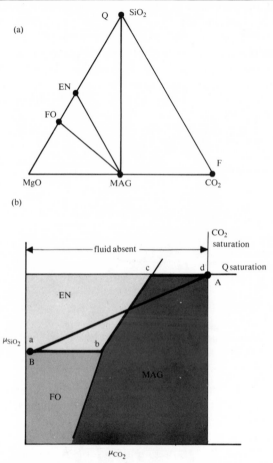

7.10 The $MgO–SiO_2–CO_2$ compatibility diagram, (a), for the diffusion-generated reaction zone problem, WE 7(c). (b) is the corresponding qualitative $\mu_{SiO_2}–\mu_{CO_2}$ diagram.

The qualitative $\mu_{CO_2}–\mu_{SiO_2}$ diagram (figure 7.10(b)) can be constructed from the compatibility diagram and Schreinemaker's rule. Note that the fields on this diagram are single-phase fields. This is because we can write a reaction between each pair of the phases which can be balanced with the end-members used on the axes of the diagram (plus any oxides whose chemical potentials are considered constant through the system—an aspect of these diagrams which is not relevant here).

If CO_2 is the only mobile oxide then the ratio of MgO to SiO_2 in the resulting reaction zones does not change. Thus the assemblages in each reaction zone are given by the intersection of a line from peridotite to CO_2 with the tie lines in the compatibility diagram. The reaction zones are peridotite, enstatite + magnesite, magnesite + quartz, then quartzite. Note that the proportions of the minerals in each reaction zone can be obtained by applying the lever rule to the appropriate tie line where the peridotite–CO_2 line intersects it. For example, the proportion of magnesite in the reaction zones increases as the proportion of forsterite in the original peridotite is increased. The reaction zone sequence can also be shown on the μ–μ diagram (abcd in figure 7.10(b)). The reaction zone sequence for the case where SiO_2 and CO_2 are equally mobile can be seen from the straight line drawn between the peridotite and the quartzite on the μ–μ diagram (ad in figure 7.10(b)). The sequence is peridotite, enstatite, magnesite and then quartzite.

References

Brady, J. B., 1977, Metasomatic zones in metamorphic rocks. *Geochim. Cosmochim. Acta*, **41**, 113–25: Source and further reading for mass transfer.

Fisher, G. W., 1977, Nonequilibrium thermodynamics in metamorphism. In Fraser D. G. (Ed.) *Thermodynamics in Geology*, Reidel, Dordrecht: Advanced reading—an alternative approach to mass transfer.

Dodson, M. H., 1976, Kinetic processes and thermal history of slowly cooled solids. *Nature*, **259**, 551–3: Advanced reading on the temperature–time dependence of rate processes.

Murr, L. E., 1975, *Interfacial Phenomena in Metals and Alloys*, Addison Wesley, New York: *and*

Christian, J. W., 1975, *The Theory of Transformations in Metals and Alloys*, Pergamon, Oxford: Very advanced reading on metallurgical aspects of processes.

Chapter 8

Processes 2

In the following sections, some of the topics from the last chapter will be combined and considered in terms of three important processes in geochemistry; buffering, metasomatism and fractionation.

Buffering

Buffering is the process by which the composition of a phase is controlled either by the compositions of the other phases during the evolution of an assemblage with changing temperature and/or pressure (internally buffered), or externally from the assemblage (externally buffered). It is assumed that equilibrium is reached more quickly than the rate of change of temperature and/or pressure with time. Written in this way, the composition of every mineral is being buffered by the compositions of the other minerals in an assemblage as the conditions are changed via the equilibrium and mass balance relations between the phases. However the term buffering is usually used in the context of the fluid phase, aqueous or silicate. In this case, external buffering can be effected by communication of the fluid in the rock with a reservoir of fluid of a particular composition in surrounding rocks, implying effective diffusion and/or infiltration. Internal buffering usually implies ineffective diffusion and/or infiltration.

Consider the metamorphic reaction (figure 8.1):

$$3CaMg(CO_3)_2 + 4SiO_2 + H_2O = Mg_3Si_4O_{10}(OH)_2 + 3CaCO_3 + 3CO_2$$

| dolomite | quartz | fluid | talc | calcite | fluid |

occurring at a particular pressure for impure limestones in the model system $CaO–MgO–SiO_2–H_2O–CO_2$ in the presence of an $H_2O–CO_2$ fluid phase.

The evolution of a calcite (CC), dolomite (DOL), quartz (Q) rock (A or B) with a particular composition fluid phase can occur in two quite

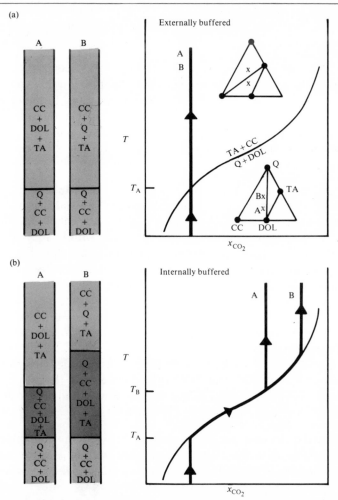

8.1 A comparison of the sequence of assemblages formed during prograde metamorphism of two limestones, A and B, which are in equilibrium with a fluid which is: (a) externally buffered; and (b) internally buffered.

different ways. Figure 8.1(a) shows the assemblages which result from the superposition of the fluid composition (or μ_{CO_2} and μ_{H_2O}) on the rock, via for example a large reservoir of fluid of that composition with which the fluid in the rock is in communication. This implies that CO_2 and H_2O are mobile. As the calcite–dolomite–quartz rock heats up, it intersects the dolomite + quartz breakdown reaction at T_A. As the composition of the fluid is fixed, the reaction goes to completion at this temperature (assuming equilibrium). Quartz and dolomite react in the ratio 4:3 (from the reaction coefficients), so that in A, quartz disappears at T_A leaving a calcite–dolomite–talc assemblage, while in B dolomite disappears leaving a quartz–calcite–talc assemblage. In a prograde sequence of metamorphosed limestones, the four-phase assemblage will be found on a line in the field if the fluid phase composition is externally controlled.

On the other hand, the assemblage may buffer the composition of the fluid phase (figure 8.1(b)). This implies that the fluid is not in communication with a large reservoir of a particular composition. If adjacent bands of limestone can be shown to have had different fluid compositions during metamorphism, this would imply further that H_2O and CO_2 were relatively immobile during metamorphism, otherwise the chemical potential gradients would have been destroyed by diffusion.

Consider rock A heating up. At T_A, dolomite + quartz start to react to give talc + calcite involving the consumption of H_2O and the evolution of CO_2. As the fluid composition is not externally buffered, the fluid composition will change as the reaction proceeds with increasing temperature. The assemblage, quartz–calcite–dolomite–talc, moves along the reaction line until one of the reactants is used up, for A this is quartz, for B dolomite. The distance moved along the lines depends on the initial proportion of reactants in the assemblage, the proportions of these reactants involved in the reaction, and the nature of the reaction. The result of buffering of the fluid phase by the assemblage during metamorphism is that the reaction assemblage, quartz–calcite–dolomite–talc, will be found over a distance in the field in a prograde sequence of metamorphic limestones. This distance will depend on the proportions of the minerals in the initial assemblage, on the initial fluid composition, the type of reaction, and on whether any reactions intersect this reaction.

It is instructive to consider the mathematics of internal buffering. Consider the effect on an assemblage which starts with $n^\circ_{CO_2}$ and $n^\circ_{H_2O}$ moles of H_2O and CO_2 in the fluid intersecting a reaction which involves y_{H_2O} and y_{CO_2} molecules of H_2O and CO_2, with increasing temperature. As the assemblage moves along the reaction the fluid contains n_{H_2O} and n_{CO_2} moles of H_2O and CO_2. If the reaction is the only force changing the composition of the fluid, then the ratio of the change in the amount of H_2O in the fluid to the change in the amount of CO_2 in the fluid will be equal to the ratio of the reaction coefficients of H_2O and CO_2, or:

$$\frac{n_{H_2O} - n^\circ_{H_2O}}{n_{CO_2} - n^\circ_{CO_2}} = \frac{y_{H_2O}}{y_{CO_2}}$$

For example, a reaction with $y_{H_2O} = 3$ and $y_{CO_2} = 1$ evolves three times as much H_2O as CO_2. This equation can be re-arranged to give:

$$n_{H_2O} = n_{H_2O}^\circ + \frac{y_{H_2O}}{y_{CO_2}}(n_{CO_2} - n_{CO_2}^\circ)$$

Or, using $A = (y_{H_2O} + y_{CO_2})/y_{CO_2}$:

$$n_{H_2O} = n_{H_2O}^\circ + (A - 1)(n_{CO_2} - n_{CO_2}^\circ) \tag{8.1}$$

Now, by definition:

$$x_{CO_2} = \frac{n_{CO_2}}{n_{CO_2} + n_{H_2O}}$$

Substituting (8.1):

$$x_{CO_2} = \frac{n_{CO_2}}{n_{CO_2} + n_{H_2O}^\circ + (A - 1)(n_{CO_2} - n_{CO_2}^\circ)}$$

Re-arranging, using $a = n_{CO_2}^\circ + n_{H_2O}^\circ$, the original number of moles of fluid:

$$n_{CO_2} = \frac{x_{CO_2}(a - An_{CO_2}^\circ)}{1 - Ax_{CO_2}}$$

We require $(n_{CO_2} - n_{CO_2}^\circ)$, the amount of CO_2 produced by the reaction, Δn_{CO_2}. Subtracting $n_{CO_2}^\circ$ from both sides and re-arranging:

$$\Delta n_{CO_2} = n_{CO_2} - n_{CO_2}^\circ = \frac{ax_{CO_2} - n_{CO_2}^\circ}{1 - Ax_{CO_2}}$$

From the definition of a, $n_{CO_2}^\circ = ax_{CO_2}^\circ$, where $x_{CO_2}^\circ$ is the initial mole fraction of CO_2. Substituting this and using Δx_{CO_2} for $x_{CO_2} - x_{CO_2}^\circ$:

$$\Delta n_{CO_2} = \frac{a}{1 - Ax_{CO_2}} \Delta x_{CO_2} \tag{8.2}$$

This is a useful expression because it relates the change in the fluid composition to the mole fractions of CO_2 in the fluid, $x_{CO_2}^\circ$ and x_{CO_2}, the reaction coefficients (via A)'and the original amount of fluid, a. The term Δn_{CO_2} is significant because it can be related to the change in the amounts of the other phases involved in the reaction due to passage along the reaction via the reaction coefficients. Thus for the dolomite + quartz breakdown reaction, the amount of talc produced, Δn_{ta} for a certain Δx_{CO_2} is $\frac{1}{3}\Delta n_{CO_2}$. The distance travelled along the reaction line (Δx_{CO_2}) to use up one of the reactants can be calculated in this way.

Δn_{CO_2} is directly proportional to the amount of fluid phase originally present in the assemblage, a. For the dolomite + quartz breakdown reaction, starting with $x_{CO_2}^\circ = 0.2$ and progressing to $x_{CO_2} = 0.5$ along the reaction:

$$\Delta n_{CO_2} = \frac{a}{1 - 2(0 \cdot 5/3)} = 0 \cdot 45 a$$

Thus, from the reaction coefficients, $\Delta n_{ta} = 0 \cdot 15a$, $\Delta n_{cc} = 0 \cdot 45a$, $\Delta n_{dol} = -0 \cdot 45a$, $\Delta n_q = -0 \cdot 60a$, and $\Delta n_{H_2O} = -0 \cdot 15a$. If the assemblage originally contained 2 mole % of fluid, we can set $a = 2$ in the above, and see that $0 \cdot 3$ mole % talc and $0 \cdot 9$ mole % calcite are formed and $0 \cdot 9$ mole % dolomite and $1 \cdot 2$ mole % quartz are used up. The important thing to note is that the amount of talc and calcite generated for $\Delta x_{CO_2} = 0 \cdot 3$ for this reaction is very small, and might be missed in thin section. If the original amount of fluid was smaller then the amount of products generated will also be smaller.

The distance travelled along the reaction line before one of the reactants is used up will also be controlled by the initial amount of fluid as well as the initial proportions of the minerals. Unless one of the reactants is originally present in very small amounts (less than a couple of per cent), the assemblage is unlikely to depart from the reaction line in the manner of figure 8.1. Instead, the assemblage will come to the intersection of the reaction line with other reactions, where another phase joins the assemblage, one of the previously formed products reacts out and the assemblage moves up a new reaction line; or, if the reaction has a maximum on a $T-x$ diagram (lines ii, iii, and iv, figure 3.9), the assemblage leaves the line near the maximum having converted reactants to products as necessary. If the assemblage is moving up a reaction line towards a maximum, the amount of reactants which have to be converted to products to continue to move up the reaction line increases dramatically towards the maximum, which means that the maximum cannot actually be reached.

Buffering is not confined to metamorphic environments. Internal buffering may operate immediately below the sediment–water interface if the sediments have a low permeability so that the aqueous solution in each small portion of sediment is effectively isolated from the next. On the other hand, if the sediments are permeable, the composition of the fluid may be externally buffered by the composition of the body of water which deposited the sediment. Buffering here involves particularly the species in solution. A transition from external buffering to internal buffering is likely to take place with burial and the onset of diagenesis.

Good examples of buffering occur in the crystallization of a large body of magma particularly if cumulus processes are operating. On initial deposition of the cumulus crystals, the intercumulus liquid is still in communication with the main body of magma and so its composition is buffered externally. With continued cumulation, the intercumulus liquid may become progressively isolated from the main body of magma, so that part of the crystallization of the intercumulus liquid takes place under internal buffering conditions. This situation gives rise to orthocumulates, while complete intercumulus crystallization under external buffering conditions gives rise to adcumulates. This situation is more complicated because the minerals are all solid solutions, so that there is a complex balance between the compositions of all the phases as cooling proceeds via the equilibrium and mass balance relations between the phases.

From the mathematics of internal buffering above, the crucial value which controls the amounts of reactants used up and the amounts of products generated for a particular change of fluid composition is the amount of the end-member originally present in the fluid. Or, put another way, the effectiveness of crystallizing a small amount of a product in changing the fluid composition depends on the amount of the end-member of interest in the fluid in the original assemblage. Thus, if we are buffering Mg^{2+} in an aqueous solution in a sediment, a really trivial amount of the product phases need to be generated to change profoundly the amount of Mg^{2+} in the solution. On the other hand, in the intercumulus environment, substantial crystallization of the product phases of a reaction buffering silica are required to change the amount of silica in the intercumulus liquid, at least initially, because the amount of silica in the liquid and the proportion of liquid to solids will be large. As crystallization proceeds the capacity of the solids to buffer the liquid increases because the proportion of liquid to solid decreases. Thus we should expect extreme fractionation in the dregs of crystallization of intercumulus liquids assuming surface equilibrium during this crystallization. For example, very small patches of quartz, alkali feldspar, amphibole and biotite are found among the main intercumulus minerals in large layered gabbros, for example Skaergaard.

Metasomatism

Metasomatism refers to the change of the composition of a rock by mass transfer, the transfer of material into and/or out of the rock. In contrast, a process which has taken place without any gain or loss of material is referred to as isochemical. Metasomatism occurs by a combination of diffusion (as a response to chemical potential gradients) and infiltration (fluid flow due to pressure and/or temperature gradients). As metasomatism actually involves changing the composition of the system, the mass transfer process must be efficient over the time which the process operates for metasomatism to occur. For diffusion, this means that the element of interest must be present in the diffusing medium, for example the fluid phase, in sufficient quantity so that the appropriate change of composition can occur in the time available, given sufficient mobility. If all elements had equivalent mobility (i.e. an atom of each element could travel unit distance under the action of a unit chemical potential gradient in the same time), then there will be effective mass transfer only of those elements with the larger solubility in the fluid phase. For infiltration, the solubility of the elements in the fluid is also important, along with the permeability of the rocks passed through and the rate of flow of the fluid.

The effect of mass transfer is obvious if diffusion and/or infiltration have not gone to completion so that a sequence of reaction zones occurs. The relative transfer between the zones can be worked out by comparing bulk analyses of the different zones assuming that the zones formed in the same composition rock. There will be little evidence in the assemblages to suggest that mass transfer has occurred if diffusion and/or infiltration has operated to the extent of removing the chemical potential gradients. A careful study might reveal a constancy of mineralogy and mineral chemistry (but not mineral proportion) if all the chemical potential gradients had been removed. A very real problem in deciding whether metasomatism has occurred, particularly on a medium or large scale, is the lack of information on the pre-metamorphic or pre-diagenetic disposition of rock compositions. For example, is the gradual change in metasedimentary rock compositions across a hill due to fluids moving through the rocks in that direction, or is it simply the reflection of some feature of the depositional environment of the original sediments? Even when metasomatism is accepted to have occurred, in for example, the hydration of ultramafic rocks to produce serpentinites, the details of the process are by no means easily resolved. If just H_2O is involved, then the rock gains volume. If the process takes place at constant volume then the addition of H_2O must be accompanied by the loss of SiO_2 and MgO. Mass transfer must be a dominant process for much smaller scale systems, of the order of the size of the grains in an assemblage, because it controls all grain growth and the development of textures. Nevertheless the details of the mass transfer cannot be resolved because we cannot tell what mineral or minerals were located where the mineral of interest now occurs, unless the process has not gone to completion.

On a large scale, the most obvious evidence for metasomatism is in the comparatively rare reaction zone sequences between, for example, granites and limestones or pelites in the contact metamorphic environment and between basic or ultrabasic rocks and metasediments in the regional metamorphic environment. Both situations might be expected to involve diffusion and infiltration. Infiltration might be particularly important in the former case, with convective fluid flow set up by the thermal effect of the granite on its environment. The chemistry and physics of such a situation is likely to be complicated as temperature and pressure gradients as well as chemical potential gradients are involved. On a small scale, the most obvious evidence for metasomatism is in the reaction textures observed in metamorphic, sedimentary and, igneous rocks, where a mineral is being replaced by another mineral of quite different composition; for example, the replacement of pyroxene by amphibole in some lavas, the replacement of aluminosilicates by muscovite in retrograde metamorphism, the alteration of feldspar to kaolinite in the weathering environment, and so on.

Given the time for mass transfer processes, the diffusional mobility of elements, and the solubility of elements in solutions, it might be expected that infiltration is a potentially more effective method for mass transfer than diffusion. This however will be the case only at low/very low pressures where rocks are sufficiently porous to allow fluid flow. Under these conditions, infiltration metasomatism may operate over great distances, even of the order of kilometres rather than the millimetres or centimetres of diffusion. This metasomatism may involve the addition of material, for example causing the cementing of sediments, as well as the removal of material, both in bulk, causing the removal of beds, and by leaching. The driving force for this fluid flow may be pressure gradients, but temperature gradients are as effective for this purpose. This has considerable relevance for ore deposition. The setting up of convective fluid flow in porous sediments by the introduction of a heat source, for example a granite, provides the driving force for mass transfer: the fluid leaches the ore materials from the large volume of rock through which the fluid passes, and then deposits them in a favourable environment.

At higher pressure, in rocks of low porosity, there will be little infiltration, even in the presence of the appropriate pressure and/or temperature gradients. Here diffusion is the main method of metasomatism. The main effects will be on a grain size scale, although mass transfer of H_2O and CO_2 into dry basic or metabasic rocks may produce reaction zones on a hand-specimen or even outcrop scale.

Fractionation

Fractionation is a process by which the composition of a system can be changed by the equilibrium formation of phases which are then chemically or physically removed from the system. Consider a new phase forming, a crystal in a cooling magma or in a prograde metamorphic rock. If diffusion in the new phase is very slow compared to the rate of growth of the phase, then the interior of the phase cannot maintain equilibrium with the rest of the assemblage, although equilibrium is maintained between the surface of the new phase and the assemblage (surface equilibrium). If the new phase forms a solid solution, then this may result in the formation of zoned grains depending on whether temperature is changing during growth, the relative amount of the new phase formed in the system and the equilibria controlling the composition of the new phase. Zoning is the best evidence of surface equilibrium, and thus of this form of fractionation.

One situation in which zoning, and thus fractionation, will occur is where surface equilibrium occurs in the heating or cooling of a system which meets a $T-x$ loop on a $T-x$ diagram. Consider figure 8.2, a $T-x$ loop in a binary system where complete solution occurs in the high and low temperature phases. If phase A of composition x is cooled from high temperature, then at T_a an infinitesimal amount of B of composition x_a is formed. With further cooling, the equilibrium composition of A and B becomes more 1-rich, while the proportion of B increases and the proportion of A decreases (figure 2.17), so that the composition of the system

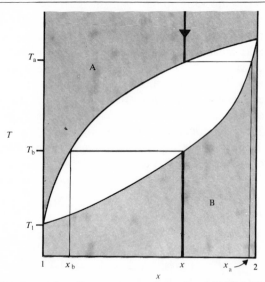

8.2 A $T-x$ diagram used to illustrate some features of fractionation in the text.

remains at x. However, if the new phase B can only maintain surface equilibrium with A, then each increment of B crystallized is removed from the equilibrating system as the next increment crystallizes. The effect is that the composition of the equilibrating system in figure 8.2 gets progressively more 1-rich with cooling. This means that coexisting compositions of A and B become progressively more 1-rich with cooling beyond the limit of equilibrium crystallization. The limit of such surface equilibrium effects is that the last dregs of phase A have the composition of pure 1, crystallize to phase B of composition of pure 1 at T_1, all the end-member 2 originally present in phase A being hidden from the equilibrating system by occurring within the zoned grains of phase B. The zoning of phase B extends from x_a to 1.

This behaviour is most familiar in igneous geology, with the $T-x$ loop a melting loop, A as the silicate liquid, and B, for example, plagioclase or olivine. Here, fractionation of the liquid is the result of surface equilibrium between plagioclase and the liquid, as well as the production of the zoned plagioclase grains which are so common in igneous rocks. This behaviour also occurs in metamorphism, the appropriate $T-x$ loop representing the equilibrium between two mineral solid solutions (e.g. figure 6.6). The zoning of metamorphic garnets can sometimes be explained in terms of surface equilibrium of the garnet with, for example, biotite in the assemblage during prograde metamorphism.

The physical removal of a phase from the system is only significant when the phase fractionated is much denser or lighter than the rest of the system. This behaviour is most familiar in the crystallization of a large body of magma. As crystals form, they sink (or rise) due to the density difference between the crystals and the magma, aided by convective overturn of the magma. The physical removal of crystals from the system has the same effect as surface equilibrium between crystals and magma in producing fractionation of the equilibrating system, in this case, the magma. This process is called crystal fractionation, fractional crystallization or just fractionation. This is in contrast to equilibrium crystallization, where the whole of the original system is maintained as the equilibrating system. Equilibrium crystallization followed by removal of the crystals changes the composition of the magma, but less effectively than fractional crystallization. Fractionation may be important during partial melting, each small increment of magma being removed from the source as it forms. This process is called fractional melting, in contrast to equilibrium (or batch) melting. This form of fractionation will occur in sedimentary and shallow diagenetic environments if a vapour is formed and if it can rise out of the system due to its lower density.

In some circumstances, the mathematics of fractionation are reasonably tractable. Consider the formation of phase B from phase A, where phase B is in surface equilibrium with phase A. Consider that there are n_A moles of phase A, and that the mole fractions of end-member 1 in phases A and B are x_{1A} and x_{1B}. A small amount of crystallization of A removes dn_A moles from the system as B. The amount of end-member 1 removed is $x_{1B} \, dn_A$, if 1 mole of A makes 1 mole of B. The change in the amount of end-member 1 in A is $d(x_{1A} n_A)$. For mass balance of end-member 1, the amount of 1 gained by B must equal the amount lost by A, or:

$$x_{1B} \, dn_A = d(x_{1A} n_A)$$

Expanding:

$$x_{1B} \, dn_A = n_A \, dx_{1A} + x_{1A} \, dn_A$$

Re-arranging gives the differential equation due to Rayleigh, whose name is often used to describe such fractionation processes (e.g. Rayleigh melting—meaning perfect fractional melting):

$$\frac{dn_A}{dx_{1A}} = \frac{n_A}{x_{1B} - x_{1A}}$$

This can be solved if x_{1B} can be expressed in terms of x_{1A}. The equilibrium relation for the reaction $(1 \text{ in } A) = (1 \text{ in } B)$ is:

$$\Delta\mu = 0 = \Delta G^\circ + RT \ln \frac{x_{1B}}{x_{1A}} + RT \ln \frac{\gamma_{1B}}{\gamma_{1A}}$$

If the activity coefficients cancel, for example if A and B are ideal solutions, then $x_{1B} = K x_{1A}$ where $K = \exp(-\Delta G^\circ / RT)$. Substituting this into the above differential equation and re-arranging:

$$(K-1)\frac{dn_A}{n_A} = \frac{dx_{1A}}{x_{1A}}$$

Integrating from $n_A = n_A^\circ$ and $x_{1A} = x_{1A}^\circ$ gives:

$$(K-1)\ln \frac{n_A}{n_A^\circ} = \ln \frac{x_{1A}}{x_{1A}^\circ}$$

Now, $F = n_A / n_A^\circ$ is the fraction of the initial amount of A available. Re-arranging gives the Rayleigh fractionation law:

$$x_{1A} = x_{1A}^\circ F^{K-1} \tag{8.3}$$

which gives the composition of A after $(1 - F)$ of A has been converted to B in terms of the initial composition of A and the equilibrium constant K. Note that x_{1A} will change dramatically with F if K is very large because F is to the power of $K - 1$. K is very large if end-member 1 greatly prefers to be in phase B, and phase A is rapidly depleted in end-member 1 by crystallizing B. K is very small if end-member 1 greatly prefers to stay in phase A, and phase B will contain little of end-member 1. These relations can be summarized on a plot of x_{1A}/x_{1A}° against F for different values of K (figure 8.3).

Although (8.3) is of interest in considering zoning and other fractionation phenomena in sedimentary and metamorphic rocks, its main realm of application comes in considering fractionation of magmas. The trace elements in magmas show a wide variety of K values for distribution between magma and different minerals. Given some idea of initial magma composition and the K values, it is possible to work out which minerals could have fractionated to produce the observed sequence of rocks. Note that the situation here is usually more complicated because more than one phase is crystallizing from the magma at one time. If more than one phase is crystallizing then (8.3) can still be used if K can be replaced by a bulk partition coefficient which reflects the contribution of each of the crystallizing phases. The same form of equation can be used to predict fractional melting relations.

Equation (8.3) applies to a grossly simplified situation of perfect fractionation (or melting) in which the phases behave in a thermodynamically ideal way. If the phases are non-ideal then an equivalent expression may be difficult to formulate and the differential equation must be solved numerically. Equations can be derived for the trapping of intercumulus liquid as a part of fractionation, and so on. Considering the crystallization of a magma, it is sufficient to realize that the two extreme ways of

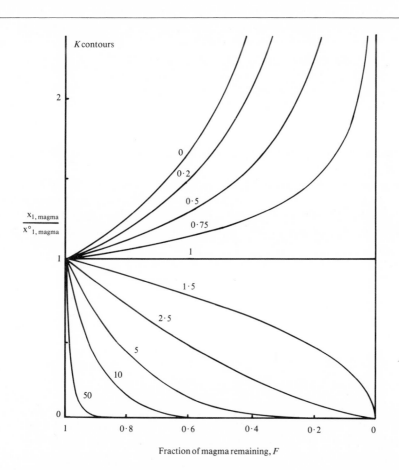

8.3 A diagram showing how $x_{1,\text{magma}}/x^{\circ}_{1,\text{magma}}$ is a function of the fraction of magma remaining, F, for particular values of K (contours). The magma becomes progressively enriched in 1 during crystallization if K is less than 1, and becomes progressively depleted in 1 during crystallization if K is greater than 1.

changing the magma composition are by equilibrium crystallization followed by magma separation, and by (perfect) fractional crystallization, which, in terms of figure 8.2, means that the magma (A) can reach x_b by equilibrium crystallization, but can reach pure 1 by fractional crystallization.

Worked examples 8

8(a) A series of impure limestones containing abundant calcite, varying amounts of dolomite and quartz and a CO_2–H_2O fluid phase with $x_{CO_2} = 0.6$ are subjected to slow heating from 500 °C to 625 °C. Show how the assemblages and the proportions of the phases change with temperature for different fluid phase behaviour, with particular reference to rocks which initially contain:

    (a) 49 mole % calcite (CC), 30 mole % dolomite (DOL), 20 mole % quartz (Q), and 1 mole % fluid (F);

    (b) 50 mole % calcite, 47 mole % dolomite, 2 mole % quartz, 1 mole % fluid.

Assume that diopside (DI) and tremolite (TR) are the only new phases involved.

Consider rock (a) using figure 6.8. If the composition of the fluid phase is externally buffered, then the assemblages developed with increasing temperature will be given by the compatibility diagrams along a vertical line at $x_{CO_2} = 0.6$ on figure 6.8. Thus the sequence of assemblages will be Q+DOL, TR+DOL, DOL+DI (all with CC). The proportions of the phases in each of the assemblages can be calculated using the lever rule or by considering the reaction which converts the assemblage into another. Using the latter approach, the mole % values can be treated as numbers of moles. The assemblage starts with 49 moles CC, 30 moles DOL, 20 moles Q and 1 mole F. On heating, when the reaction line is intersected at 530 °C, one of the reactant phases must be used up before the assemblage can leave the line. As 8Q are used up for 5DOL and as there is less Q than DOL in the assemblage, Q will be the phase used up. The reaction:

$$5DOL + 8Q + H_2O = TR + 3CC + 7CO_2 \quad [DI]$$

must take place $2\frac{1}{2}$ times to use up all the Q, resulting in the assemblage, 56·5 moles CC, 17·5 moles DOL, 2·5 moles TR with a fluid which has changed by +17·5 moles of CO_2 and −2·5 moles of H_2O. This effect on the fluid must be countered by the gain of some H_2O and a loss of most of this CO_2 via the external reservoir which is controlling the fluid composition. It would seem reasonable to expect that the amount of fluid in the rock remains small, and a constant value of 1 mole % is assumed in the calculations in this worked example.

On further heating, the assemblage reaches the [Q] reaction at 610 °C:

$$TR + 3CC = 4DI + DOL + CO_2 + H_2O \quad [Q]$$

Note that the assemblage does not 'see' the [DOL] reaction because it is not appropriate to this bulk composition, it only affects compositions with the Q+TR(+CC+F) assemblage. At [Q], the tremolite is the first reactant to disappear, leaving the assemblage with 49 moles CC, 20 moles DOL and 10 moles DI, involving the application of [Q] 2·5 times.

The way the proportions change with temperature can be plotted on diagrams like figure 8.4. From a practical point of view of identifying temperature markers (isograds) in a prograde sequence of metamorphic limestones, it is useful to plot such diagrams on a volume % basis because we observe volume proportions not mole proportions in hand specimen and thin section. The TR-bearing assemblage is converted to a volume % basis:

Phase	Moles	(mole %)	$\times$ MV	Volume %
CC	56·5	73·9	208·65	53·6
DOL	17·5	22·9	112·60	28·9
TR	2·5	3·2	68·23	17·5

Note the dramatic increase in the percentage of TR from the mole to the volume basis due to the large molar volume of TR. On a volume % basis, the sequence of assemblages is:

CC 43·2%   DOL 46·0%   Q 10·8%
CC 53·6%   DOL 28·9%   TR 17·5%
CC 48·2%   DOL 34·3%   DI 17·5%

This is drawn up in figure 8.4. The appearance of TR and later DI as well as the disappearance of Q and TR takes place suddenly across the appropriate reactions. Such behaviour would provide excellent isograds, for example the appearance of TR with the concomitant disappearance of quartz *or* dolomite, although the temperature of each isograd would depend on the composition of the fluid in the reservoir.

If however the fluid phase composition is buffered by the assemblage during prograde metamorphism, then the sequence of assemblages is much more complicated. On being heated, rock (a) intersects [D1] at about 530 °C. In starting to convert DOL+Q to TR+CC, H_2O is used up and CO_2 generated, and the assemblage moves along [DI] towards I in figure 6.8 converting reactants to products. These amounts can be calculated using (8.2). For DI, $A = \frac{6}{7}$ and using $a = 1$ and $x^{\circ}_{CO_2} = 0·6$:

$$\Delta n_{CO_2} = \frac{x_{CO_2} - 0·6}{1 - \frac{6}{7}x_{CO_2}}$$

Moving to I involves $x_{CO_2} = 0·93$ and $\Delta n_{CO_2} = 1·63$. Thus, from the reaction coefficients for [DI], 0·23 moles of TR and 0·7 moles of CC are generated while 1·86 moles of Q and 1·16 moles of DOL are consumed.

What happens at I? At I, the assemblage is joined by DI and concurrently the assemblage loses one of the previously formed phases. The reaction taking place at I cannot involve any change of fluid composition, so the reaction must involve a fluid of composition $x_{CO_2} = 0 \cdot 93$. As we are considering CC to be present in excess, we can balance the reaction for $CaCO_3$ after finding the other reaction coefficients, in the way we did previously for reactions involving H_2O and CO_2 in excess. If we treat CC in this way, we should not include an equation for balancing CaO as the amount of CC to be included later will accomplish this. The compositions of the phases can be represented thus:

	MgO	SiO_2	CO_2	H_2O	$CaCO_3$
$CaMg(CO_3)_2$	1	0	1	0	1
$Ca_2Mg_5Si_8O_{22}(OH)_2$	5	8	−2	1	2
$CaMgSi_2O_6$	1	2	−1	0	1
SiO_2	0	1	0	0	0
Fluid (F′)	0	0	0·93	0·07	0

Noting that, for example, there are $-2CO_2$ in TR if $CaCO_3$ is used to represent the CaO in TR. The equations to solve to find the reaction coefficients, with the reaction coefficient for TR = 1, are:

for	MgO	(a)	DOL + 5	+ DI	= 0
	SiO_2	(b)	8 + Q + 2DI		= 0
	H_2O	(c)	1	+ 0·07F′ = 0	
	CO_2	(d)	DOL − 2	− DI + 0·93F′ = 0	

From (c), F′ = −14·29. Substituting F′ into (d) and then subtracting the result from (a) gives DI = −10·14. Substituting DI into (a) gives DOL = 5·14. Substituting DI into (b) gives Q = 12·28. Balancing for CaO gives CC = 3. The reaction can be written as:

$$12 \cdot 28Q + 5 \cdot 14DOL + 3 \cdot 00CC + 1 \cdot 00TR = 10 \cdot 14DI + 14 \cdot 29F'$$

If this reaction takes place 0·23 times then all the TR is used up, resulting in 2·33 moles of DI being formed and 2·82 moles of Q, 1·18 moles of DOL and 0·69 moles of CC used up. The assemblage is now CC + DOL + Q + DI and so it moves along [TR] converting DOL + Q to DI. Now, [TR] contains a maximum on a T–x diagram at $x_{CO_2} = 1$. As mentioned in the text, the conversion of reactants to products accelerates as the maximum is approached. This can be seen by applying (8.2) to [TR]. For this reaction, $A = 1$, and, if fluid loss from the assemblage is a continuous process, we can continue to take $a = 1$, along with $x^\circ_{CO_2} = 0 \cdot 93$ so:

$$\Delta n_{CO_2} = \frac{x_{CO_2} - 0 \cdot 93}{1 - x_{CO_2}}$$

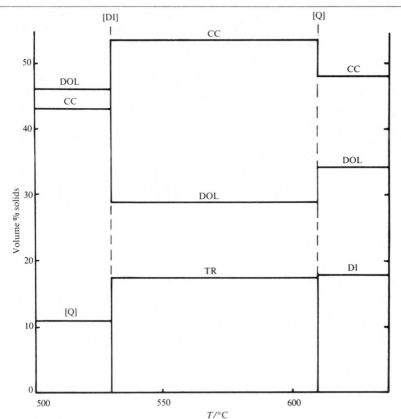

8.4   Volume proportions as a function of temperature for rock (a) in WE 8(a) for a fluid phase whose composition is externally buffered.

The acceleration is due to $\Delta n_{CO_2} \to \infty$ as $x_{CO_2} \to 1$, because of $(1 - x_{CO_2})$ in the denominator of the expression. The assemblage contains 49·01 moles of CC, 26·66 moles of DOL, 15·31 moles of Q and 2·33 moles of DI on leaving I. The reaction coefficients for [TR] mean that Q is used up twice as quickly as DOL so that the assemblage will leave [TR] when all the Q is used up. By $x_{CO_2} = 0.95$, 0·4 moles of Q are used up; by $x_{CO_2} = 0.97$, 1·33 moles of Q; by $x_{CO_2} = 0.99$, 6 moles of Q; by $x_{CO_2} = 0.996$, 15·31 moles of Q are used up. The assemblage now contains 49 moles of CC, 20 moles of DOL and 10 moles of DI, and will continue to higher temperature with a fluid of composition $x_{CO_2} = 0.996$.

The numbers of moles at each stage of this calculation can be converted to a volume % basis giving:

	43·15% CC	41·03% DOL	10·82% Q	
arrive I	44·06% CC	44·55% DOL	9·88% Q	1·51% TR
leave I	44·24% CC	43·50% DOL	8·49% Q	3·76% DI
	48·17% CC	34·26% DOL	17·57% DI	

This is shown in figure 8.5. Note that the small amount of TR developed in the assemblage in passing along [DI] might be missed in thin section, while the appearance of DI either at I, or soon after, in passing along [TR] would not be missed and would make a good isograd as the temperature of the appearance of DI will not be strongly dependent on the initial proportions of dolomite and quartz or the initial fluid composition.

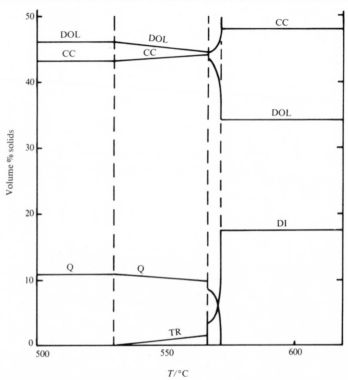

8.5   Volume proportions as a function of temperature for rock (a) in WE 8(a) for a fluid phase whose composition is internally buffered.

For rock (b), exactly the same procedures can be followed as in the above. As the amount of Q in the original assemblage is so small, the amount of products produced in reactions which use up Q is small. For externally buffered fluid compositions, the volume % of the minerals in the successive assemblages are given in figure 8.6.

For the fluid phase buffered by the assemblage, the assemblage on arrival at I contains 50·70 moles of CC, 45·84 moles of DOL, 0·14 moles of Q and 0·23 moles of TR. Looking at the reaction taking place at I, Q will be used up before TR for this rock. Therefore this rock departs from I along [Q], converting TR + CC to DOL + DI. As the assemblage only contains

0·22 moles of TR, this is used up relatively quickly, and the assemblage goes to higher temperatures with a fluid phase of composition $x_{CO_2} = 0·80$. This is summarized in figure 8.7. The amounts of Q, DI and TR involved might be missed in thin section.

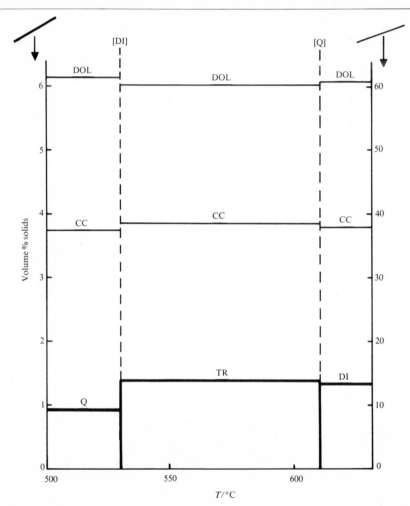

8.6 Volume proportions as a function of temperature for rock (b) in WE 8(a) for a fluid whose composition is externally buffered. Note the different scales on the two vertical axes.

Some generalizations can be made about the prograde metamorphism of limestones which have excess calcite and variable amounts of dolomite and quartz with the fluid phase composition buffered by the assemblage during metamorphism (figure 8.8). Unless the amounts of Q or DOL in the original rock are trivial, then the assemblage will pass along [DI] to I then along [TR] where diopside will be formed in a substantial amount over a small temperature interval, the amount depending on the initial abundances of Q and DOL. For trivial amounts of Q, larger than 1·86 moles, the assemblage will pass along [DI] to I, lose Q at I, and pass along [Q] until all the TR is used up. If there is initially less than this

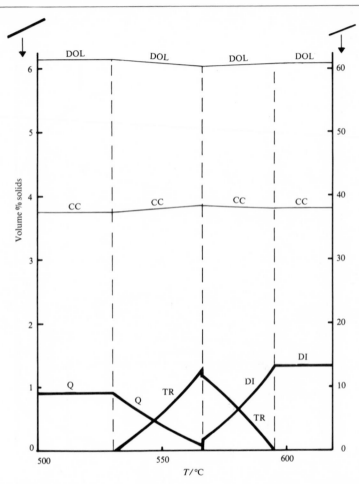

8.7   Volume proportions as a function of temperature for rock (b) in WE 8(a) for a fluid whose composition is internally buffered.

amount of Q, then the assemblage will not reach I, but will leave [TR] and heat up at constant x_{CO_2} until intersecting [Q], where the small amount of TR formed along [DI] will be converted to DI. For trivial amounts of dolomite, greater than 1·16 moles, the assemblage will pass along [DI] to I, lose DOL at I, and then pass along [DOL] until the TR is converted to DI. For smaller amounts of DOL, the assemblage will not

reach I, and will move up from [DI] to [DOL], where the small amount of
TR formed on [DI] will be converted to DI. From an isograd point of
view, substantial diopside is only formed on [TR] in these limestones, if
buffering of the fluid phase by the assemblage operates during prograde
metamorphism. TR should be little developed, if buffering along [DI] is
the only reaction responsible for its appearance in the assemblage. On the

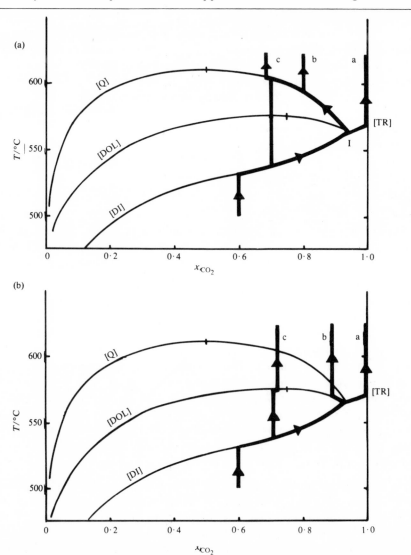

8.8 Summary of paths of rocks which initially have excess calcite and a fluid with
$x_{CO_2} = 0.6$; and different proportions of quartz, (a), 20 mole % in a, 2 mole % in
b, and 0·3 mole % for c; and different proportions of dolomite, (b), 15 mole %
for a, 2 mole % for b, and 0·2 mole % for c. (WE 8(a))

other hand, if the fluid phase composition was externally buffered during metamorphism, then the appearance of new minerals would occur abruptly and at different temperatures depending on fluid phase composition. Tremolite would be much more abundant in a prograde sequence of metamorphic rocks. Diopside will appear by the breakdown of TR + Q or TR + CC rather than by [TR], unless the fluid composition is very CO_2-rich ($x_{CO_2} > 0.93$).

Rock sequences in which the fluid phase was internally buffered during metamorphism could be distinguished easily from sequences metamorphosed under a constant fluid composition if either of these two extremes of behaviour were the only possibilities. If the communication with an external reservoir was feeble, then some combination of the above would result. Similarly fluid behaviour might change during metamorphism, with, for example, improved diffusion of H_2O and CO_2 at higher temperatures. Clearly there are many possibilities even in this simple system. Nevertheless it is important to identify the likely sequences of assemblages for different fluid phase behaviour, particularly if the presence or absence of minerals is to be used to indicate the conditions of formation of assemblages, or as isograds in the field.

8(b) Consider the development of diffusion metasomatic reaction zones between a limestone containing calcite and dolomite with a CO_2 fluid phase and a quartzite with a H_2O fluid phase, at 4 kbar and 525 °C, if H_2O, CO_2 and SiO_2 are mobile.

By analogy with the previous worked example on this type of problem, a $\mu - \mu - \mu$ block diagram is needed to represent the mineralogies in the reaction zones. First we can look at a $T - x_{CO_2}$ diagram for quartz-present limestones (figure 8.9(a)). This is the same diagram as figure 6.8, but as Q is in excess rather than CC, the [Q] reaction is not required and [TR] becomes [TR, CC] as this reaction is degenerate (see p. 141). The compatibility diagrams are now projections from Q, CO_2 and H_2O in CaO–MgO–SiO_2–H_2O–CO_2. The positions of these reactions will move to higher temperatures as the chemical potential of silica is reduced below the level required for Q to be present. Under these conditions, the fluid will not be saturated with SiO_2.

It is worthwhile considering the correspondence between this $T - x$ diagram and the $\mu_{CO_2} - \mu_{H_2O}$ diagram involving the same reactions. Although this can be done quantitatively using the same data used to construct the $T - x$ diagram, it is sufficient to draw the diagram qualitatively, using the slopes of the reactions determined from the reaction coefficients (figure 8.9(b)). If the diagram had been drawn quantitatively then the positions, but not the slopes, of the reaction lines would change with changing temperature and pressure. For a particular temperature and pressure, a line, ab, can be drawn across the diagram corresponding to the presence of a H_2O–CO_2 fluid phase. Within this line is a region characterized either by fluid-absent conditions or by the H_2O and CO_2 of the fluid phase being diluted by another fluid end-member (for example, CH_4). The region outside the line ab is physically inaccessible because it implies $P_{H_2O} + P_{CO_2}$ greater than the load pressure on the rock. This diagram is

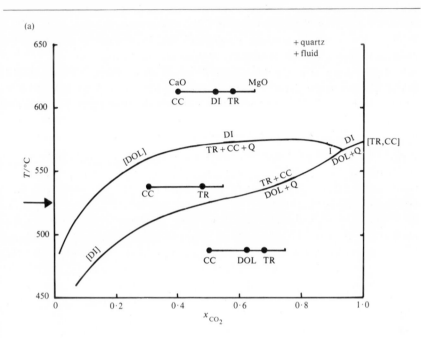

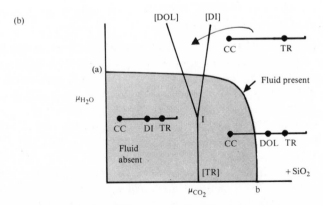

8.9 A T–x_{CO_2} diagram, (a), for quartz-excess rocks in the system CaO–MgO–SiO_2–H_2O–CO_2 in the presence of an H_2O–CO_2 fluid phase at 4 kbar. (b) is a qualitative μ_{H_2O}–μ_{CO_2} diagram for 4 kbar and 525 °C showing the fluid-present line a–b and the fluid-absent region (shaded). (**WE 8(b)**)

'+SiO_2' or more generally, it is a constant μ_{SiO_2} section of the $\mu-\mu-\mu$ diagram we are going to construct.

Metasomatic zoning due to H_2O and CO_2 mobility can be seen on such a diagram. A Q+DI+CC rock with an H_2O fluid phase (at a) against a Q+DOL+CC rock (at b) will develop a Q+CC+TR reaction zone as both CO_2 and H_2O diffuse between the two rocks. Each rock will maintain its original proportions of CaO–MgO–SiO_2 as none of these are mobile. The original boundary between the rocks can be located by a change in the proportions of CC, Q and TR within the reaction zone, if the original rock compositions were different with respect to CaO–MgO–SiO_2. The diffusion 'path' in the diagram is along the fluid-present line, rather than the straight line between a and b because there is no way in which a fluid-absent reaction zone (implied by a straight line path) could develop by diffusion from rocks which originally had a fluid phase.

Returning to the problem, the next stage is to draw equivalent μ_{H_2O}–μ_{SiO_2} and μ_{CO_2}–μ_{SiO_2} diagrams (figures 8.10(a) and (b)). These are at constant μ_{CO_2} and μ_{H_2O} respectively. Both will have a quartz-saturation line, corresponding to the appearance of Q in assemblages. Inside this line the assemblages are Q-absent. The three $\mu-\mu$ diagrams can now be assembled as a $\mu-\mu-\mu$ block diagram (figure 8.11). The Y-shape of the reactions in figure 8.9(b) extends backwards into the block diagram. The 4 kbar and 525 °C conditions constrain the relative position of the Y to the fluid-present surface (shaded) at the Q-saturation surface (back plane of block) to correspond to figure 8.9(a) at these conditions. The intersection of this solid Y with the fluid-present surface gives the relations of interest in this problem. A corresponds to the quartzite, with quartz present and a H_2O fluid phase. B corresponds to the limestone, the position of B on ab reflecting the chemical potential of SiO_2 in the assemblage (i.e. the amount of SiO_2 in the CO_2 fluid phase).

The diffusion 'path' for mobile CO_2, H_2O and SiO_2 will be some line connecting A and B on the fluid-present surface on figure 8.11. Therefore the zone sequence is either limestone, CC+DI, quartzite; or limestone, CC+TR, CC+DI, quartzite, with all the reaction zones formed in the limestone, and all the reaction zones having the same CaO/MgO ratio as the original limestone. The second reaction zone sequence makes sense on figure 8.9, in terms of a horizontal line at 525 °C. The first sequence makes sense if [DOL] and [DI] actually intersect again at low temperature and x_{CO_2}, with [TR, CC] re-appearing as a stable reaction. Thus substantial reduction of the activity of SiO_2 will bring this low temperature stable part of [TR, CC] up to 525 °C and so will prevent the appearance of a TR-bearing reaction zone.

8(c) Consider fractionation in the binary $NaAlSi_3O_8$–$CaAl_2Si_2O_8$ system at 1 bar with particular reference to a liquid of initial composition

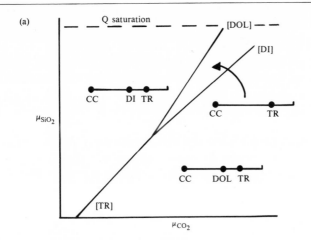

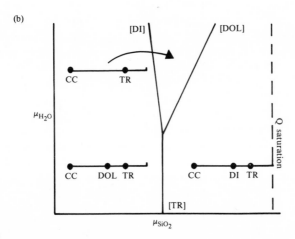

8.10   A qualitative μ_{SiO_2}–μ_{CO_2} diagram, (a), and a qualitative μ_{H_2O}–μ_{SiO_2} diagram, (b), for CC–DOL–DI–TR–Q–F equilibria. (WE 8(b))

$x_{CaAl_2Si_2O_8, liquid} = 0.5$, using

$\Delta G° = 120.6 - 0.0661\,T$ kJ   for   anorthite (solid)
$\quad\quad = $ anorthite (liquid)

$\Delta G° = \;\; 54.9 - 0.0395\,T$ kJ   for   albite (solid)
$\quad\quad = $ albite (liquid)

assuming that the plagioclase solid solution and the liquid solution are both ideal molecular solutions.

Using subscript Na for $NaAlSi_3O_8$ and Ca for $CaAl_2Si_2O_8$, and S for plagioclase and L for liquid, the equilibrium relations for the two reactions are:

$$\Delta\mu = 0 = 120\cdot6 - 0\cdot0661\,T + RT\ln\frac{x_{Ca,L}}{x_{Ca,S}}$$

$$\Delta\mu = 0 = 54\cdot9 - 0\cdot0395\,T + RT\ln\frac{x_{Na,L}}{x_{Na,S}}$$

Re-arranging:

$$x_{Ca,S} = x_{Ca,L}\exp\left(\frac{14500}{T} - 7\cdot95\right) = x_{Ca,L}K_{Ca}$$

$$x_{Na,S} = x_{Na,L}\exp\left(\frac{6600}{T} - 4\cdot75\right) = x_{Na,L}K_{Na}$$

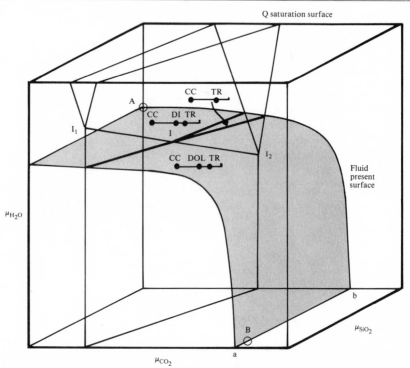

8.11 A qualitative μ_{SiO_2}–μ_{H_2O}–μ_{CO_2} diagram showing the intersection of the fluid-present surface (shaded) with the three reactions. The quartzite plots at A, the limestone at B. (WE 8(b))

Now, as we are considering the binary system, $x_{Na,S} = 1 - x_{Ca,S}$ and $x_{Na,L} = 1 - x_{Ca,L}$. Making these substitutions into the second equation, and then adding the resulting equation to the first:

$$1 = x_{Ca,L}K_{Ca} + (1 - x_{Ca,L})K_{Na}$$

Re-arranging:

$$x_{Ca,L} = \frac{1 - K_{Na}}{K_{Ca} - K_{Na}}$$

Substituting this in $x_{Ca,S} = x_{Ca,L} K_{Ca}$:

$$x_{Ca,S} = \frac{K_{Ca}(1 - K_{Na})}{K_{Ca} - K_{Na}}$$

These two equations allow the $T–x$ diagram for this system to be plotted, figure 8.12. For example, at 1623 K, $K_{Ca} = 2·675$ and $K_{Na} = 0·505$, so $x_{Ca,S} = 0·61$ and $x_{Ca,L} = 0·23$, by substitution into these equations.

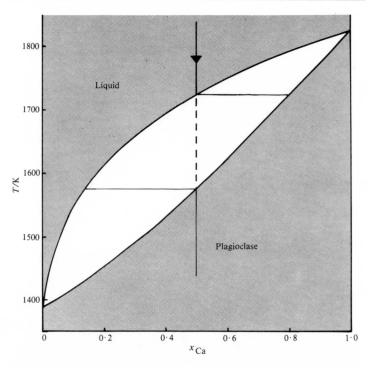

8.12   A $T–x$ diagram for plagioclase–liquid equilibria in $NaAlSi_3O_8$–$CaAl_2Si_2O_8$ calculated from the thermodynamic data in WE 8(c), used to illustrate fractionation in this system.

From the discussions in the text, we can see from the diagram that for:

(a) Equilibrium: continued crystallization of plagioclase (which changes composition from $x_{Ca} = 0·8$ to $0·5$) is accompanied by the liquid changing composition from $x_{Ca} = 0·5$ to $0·14$, when the last dregs of liquid disappear.

(b) Fractional crystallization: the liquid changes composition from $x_{Ca} = 0.5$ to 0.0 as each increment of plagioclase which crystallizes is removed from the equilibrating system, either by sinking (or rising) or by surface equilibrium being maintained between the plagioclase and the liquid.

It is interesting to compare how these two cases differ from the point of view of fraction of liquid, F, remaining at each stage during cooling, and the compositions of the phases at each stage. Using superscript $^\circ$ for initial composition, a straightforward mass balance relation for Ca gives a relationship between F, $x_{Ca,L}$ and $x_{Ca,S}$:

$$Fx_{Ca,L} + (1-F)x_{Ca,S} = x^\circ_{Ca,L}$$

Re-arranging:

$$F = \frac{x^\circ_{Ca,L} - x_{Ca,S}}{x_{Ca,L} - x_{Ca,S}}$$

Each x_{Ca} value can be expressed in terms of K's using the above equations, thus, with some re-arranging:

$$F = \frac{x^\circ_{Ca,L}\left(\dfrac{K_{Ca} - K_{Na}}{1 - K_{Na}}\right) - K_{Ca}}{1 - K_{Ca}}$$

Substitution into this equation with $x^\circ_{Ca,L} = 0.5$ at each temperature gives the broken line in figure 8.13.

From the Rayleigh fractionation equation (8.3):

$$x_{Ca,L} = x^\circ_{Ca,L}F^{K_{Ca}-1}$$

$$x_{Na,L} = x^\circ_{Na,L}F^{K_{Na}-1}$$

Substituting $x_{Na,L} = 1 - x_{Ca,L}$ into the second equation, and then adding this equation to the first:

$$1 = x^\circ_{Ca,L}F^{K_{Ca}-1} + x^\circ_{Na,L}F^{K_{Na}-1}$$

Substituting $x^\circ_{Ca,L} = x^\circ_{Na,L} = 0.5$ and re-arranging:

$$0 = F^{K_{Ca}-1} + F^{K_{Na}-1} - 2$$

This equation has to be solved iteratively as it is not possible to express T in terms of F. At 1623 K, $K_{Ca} = 2.675$ and $K_{Na} = 0.505$, so the equation to solve is:

$$0 = F^{1.675} + F^{-0.495} - 2 = f$$

The procedure is to guess F and evaluate this expression, f, making new estimates until the expression is equal to zero, $f = 0$. For $F = 0.2$, $f = 0.286$; for $F = 0.4$, $f = -0.211$. Thus, the correct value of F must lie between 0.2 and 0.4, as the value of f has gone from positive, through 0, to negative between these values. We can either guess the next value to try, or we can linearly interpolate between the values (i.e. assume that the expression, f, is linear in F in this range of F). Using the expression for a straight line:

$$\frac{f-0\cdot286}{-0\cdot211-0\cdot286}=\frac{F-0\cdot2}{0\cdot4-0\cdot2}$$

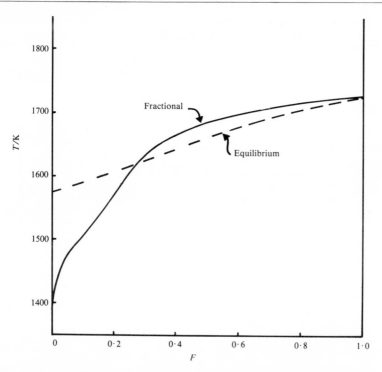

8.13   A diagram showing how the fraction of plagioclase liquid remaining, F, decreases with cooling for equilibrium (broken line) and fractional crystallization (full line) for an initial liquid composition of $x_{\mathrm{CaAl_2Si_2O_8,liquid}}=0\cdot5$. (WE 8(c))

The required value of F is for $f=0$, so solving this for $f=0$ gives $F=0\cdot315$. This linear interpolation expression can be written generally, if f_1 refers to F_1 and f_2 refers to F_2, then the required value of F is:

$$F=F_1-f_1\frac{F_2-F_1}{f_2-f_1}$$

Trying $F=0\cdot315$, the expression, $f=-0\cdot084$. Applying the linear interpolation equation to $F_1=0\cdot2$ and $F_2=0\cdot315$, then the new value of F is $0\cdot289$, for which $f=-0\cdot026$. As the iteration is repeated the value of f

approaches 0. We only need F to two decimal places, so one further iteration with $F_1 = 0.315$ and $F_2 = 0.289$, giving $F = 0.277$, with $f = 0.004$, is adequate for our purposes. Thus at 1623 K, $F = 0.28$. Such iterative calculations are tedious by hand, but are easy on a simple programmable calculator.

The iterations were repeated for a range of temperatures and the results plotted on figure 8.13 as the full line. There is relatively little difference between the curves for equilibrium and fractional crystallization until more than $\frac{2}{3}$ of the liquid has crystallized. Comparing figures 8.12 and 8.13 at each temperature allows the construction of figure 8.14, which shows

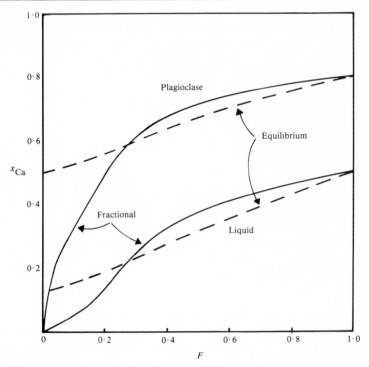

8.14 A diagram showing how the fraction of liquid remaining is related to the liquid and solid compositions for equilibrium (broken lines) and fractional crystallization (full lines). (WE 8(c))

how the compositions of the liquid and solid phases change with progressive crystallization of the liquid. Note again that there is little difference between equilibrium and fractional crystallization until more than $\frac{2}{3}$ of the liquid has crystallized. For example, very substantial fractional crystallization of such a liquid is required to effect pure albite crystallization; more than 98% of the original liquid must be crystallized.

Problems 8

8(a)   Use the phase diagrams in WE 6(i) and the buffering equations to:

(1)   Find the sequence of assemblages developed from a $CH+BR$ assemblage heated from 380 °C and $x_{CO_2} = 0.0002$ for internal and for external buffering of the fluid phase.

(2)   Find the sequences of assemblages developed from a $CH+BR+MAG$ assemblage heated from 380 °C and $x_{CO_2} = 0.0003$ for internal and for external buffering of the fluid phase.

References

Greenwood, H. J., 1975, Buffering of pore fluids by metamorphic reactions. *Am. J. Sci.*, **275**, 573–93: Source and further reading on the mathematics of buffering.

Brady, J. B., 1977, Metasomatic zones in metamorphic rocks. *Geochim. Cosmochim. Acta*, **41**, 113–25: Source and further reading on metasomatism.

Korzhinskii, D. S., 1970 *Theory of Metasomatic Zoning*, Oxford University Press, London: Very advanced text on metasomatism, particularly infiltration metasomatism.

Wood, B. J. and Fraser, D. G., 1976, *Elementary Thermodynamics for Geologists*, Oxford University Press, London: Chapter 6 is further reading on fractionation—with particular reference to trace element behaviour.

Allegre, C. J. et al., 1977, *Systematic use of trace elements in igneous processes. Contr. Mineral. Petrol.*, **60**, 57–75: Advanced reading on trace element fractionation.

Chapter 9
Petrological Topics

This chapter is a series of topics across the geological field in the form of worked examples, illustrating ideas about geological processes which can be obtained by mainly qualitative equilibrium considerations of rocks or groups of rocks.

Worked examples 9

9(a) Schematic P–T diagrams for ultrabasic (mantle composition) and basic (silica-saturated basalt) rocks are given in figures 9.1(a) and (b). Combine the two phase diagrams for the system CaO–MgO–Al$_2$O$_3$–SiO$_2$ (CMAS) with clinopyroxene always present, ignoring solid solutions in order to find the assemblages developed in other basic and intermediate composition igneous rocks at high temperatures and pressures.

Figure 9.1(a) is the schematic phase diagram for mantle composition ultrabasic rocks, while figure 9.1(b) is the schematic phase diagram for silica-saturated basalt. The diagrams cover the P–T range for the crust and the upper mantle, and so are particularly applicable to the mineralogy of the lower crust and upper mantle. The dispositions of the lines in the two diagrams are unrelated as they refer to different composition systems. It is useful to combine the two diagrams to see which reactions affect other compositions and to see what assemblages are obtained.

The compositions can be modelled in CaO–MgO–Al$_2$O$_3$–SiO$_2$ (CMAS) if we assume that the other oxides, particularly FeO and Na_2O, do not affect the mineralogy. Approximate phase relations can be unravelled by ignoring these other oxides, even though, for example, the width of the garnet granulite field in figure 9.1(b) depends on the amount of Na_2O in the system. One cause of considerable complexity in the phase diagrams for CMAS is the substantial range of solid solution shown by the

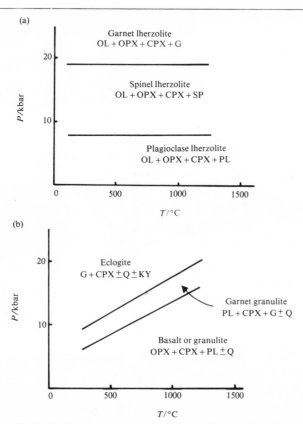

9.1 Schematic $P–T$ diagrams for mantle compositions, (a), and quartz-saturated basalts, (b). (WE 9(a))

pyroxenes and garnets. Again, approximate phase relations can be shown by using $CaMgSi_2O_6$ for clinopyroxene (CPX), $MgSiO_3$ for orthopyroxene (OPX), $Mg_3Al_2Si_3O_{12}$ for garnet (G), $CaAl_2Si_2O_8$ for plagioclase (PL), Mg_2SiO_4 for olivine (OL), Al_2SiO_5 for kyanite (KY), $MgAl_2O_4$ for spinel (SP) and SiO_2 for quartz (Q). The next problem is that compatibility diagrams for CMAS are tetrahedra, which are not easy to draw, and less easy to look at (figure 9.2(a)). The way round this is to use a projection from a phase always present in the assemblages of interest. Clinopyroxene is used because it is present in many of the assemblages of interest, even though clinopyroxene does show a wide range of compositions in natural assemblages. This means that the projected position of a phase depends on the composition of the clinopyroxene from which it is projected. Nevertheless a qualitative P–T diagram can be constructed assuming that the projection clinopyroxene is $CaMgSi_2O_6$. The phases are projected from CPX onto the plane, $CaAl_2O_4$–MgO–SiO_2(figure 9.2(b)); some of the construction lines are shown on figure 9.2(a).

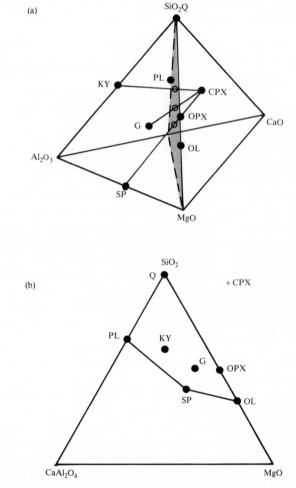

9.2 The CMAS compatibility tetrahedron, (a), with the positions of the phases indicated. The plane of projection is shaded. Open circles indicate the positions of phases which have to be projected from CPX onto this plane. (b) is the projection from CPX onto $CaAl_2O_4$–MgO–SiO_2 (using an equilateral triangle for this plane). (WE 9(a))

Plotting those phases already on the plane presents no problem. Plotting $Mg_3Al_2Si_3O_{12}$ on the plane involves expressing this formula in terms of the compounds at the apices of the diagram, $CaAl_2O_4$, SiO_2, and MgO, plus the excess phase $CaMgSi_2O_6$. Thus, by inspection:

$$Mg_3Al_2Si_3O_{12} = CaAl_2O_4 - CaMgSi_2O_6 + 4MgO + 5SiO_2$$
$$MgAl_2O_4 = CaAl_2O_4 - CaMgSi_2O_6 + 2MgO + 2SiO_2$$
$$Al_2SiO_5 = CaAl_2O_4 - CaMgSi_2O_6 + MgO + 3SiO_2$$

Thus, $MgAl_2O_4$ lies on the intersection of the $MgO/SiO_2 = 1$ and $MgO/CaAl_2O_4 = 2$ lines.

The tie lines on figure 9.3(a) are appropriate to the crystallization of igneous rocks at the Earth's surface. The shaded area represents the compositions of common igneous rocks, in terms of the assemblages $Q + PL + OPX + CPX$ and $OL + PL + OPX + CPX$. The compatibility diagrams at each $P-T$ can be used to find the assemblage for these rocks if metamorphosed at those conditions.

Figure 9.3(b) shows the lines from figures 9.1(a) and (b) superimposed on the same $P-T$ diagram. First the reactions are labelled using the difference in the assemblages across the lines in figure 9.1. For example, in going from the spinel lherzolite to the garnet lherzolite field, involves replacing a tie line between SP and OPX by a tie line between G and OL. Then the fields between the lines can be labelled with the appropriate compatibility diagrams. A new reaction, $PL + OPX + SP = G$, must be inserted between $Q + G = OPX + PL$ and $SP + OPX = PL + OL$ to make the sequence of compatibility diagrams consistent from high to low pressure. This new reaction will intersect $Q + G = OPX + PL$ at some low temperature at A, giving rise to a number of new reactions, one of which will intersect $KY = G + PL + Q$ at B generating more reactions. This part of the diagram is completely schematic.

One way of realizing the significance of this diagram is to redraw it and mark on those lines which apply to particular rock compositions. Clearly figures 9.1(a) and (b) are just these sort of diagrams for the mantle composition and silica-saturated basalts. A picritic basalt whose composition plots within the G–OPX–OL triangle in figure 9.3(a) 'sees' the same reactions as the mantle composition, and so figure 9.1(a) applies. The intermediate pressure assemblage will be $PL + SP + OPX + CPX$ if the composition lies within SP–PL–OPX rather than SP–OPX–OL. Figures 9.4(a)–(c) apply to olivine basalts; which one of these applies depending essentially on the amount of olivine originally present. Diagram (a) applies to compositions which plot in the area of intersection of triangles OL–OPX–G and PL–OPX–SP; (b) to the intersection of Q–OPX–G and PL–OPX–SP; and (c) to the intersection of KY–G–Q and PL–OPX–SP. Note how the relevant reactions depend on the composition of the system. Note also that the olivine basalts represented by diagrams (b) and (c) develop Q-bearing assemblages at high pressure although the rocks as lavas would contain OL. Diagram (d) applies to more Q- and PL-rich compositions; the high pressure assemblage contains PL rather than G, compared with the silica-saturated basalts.

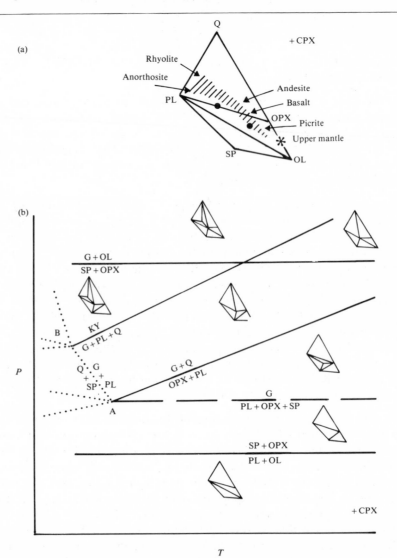

9.3 The CPX-projection compatibility diagram at low pressure, (a), showing the normal range of composition of igneous rocks, and the composition of the upper mantle. (b) is the qualitative $P–T$ diagram produced by combining figures 9.2(a) and (b). Broken and dotted lines are conjectural. (WE 9(a))

There are good possibilities of locating a set of high pressure rocks, eclogites or granulites, on a *P–T* diagram, if there is some variation in the compositions of the rocks. For example, if a series of basalts from picrites to quartz-tholeiites, had been metamorphosed at high *P–T* conditions, the different assemblages developed in the different compositions plus figure 9.3 or 9.4 would allow the appropriate *P–T* region to be located.

It is worthwhile bearing in mind that certain of the features in figure 9.4 might not occur in natural compositions because of the assumptions

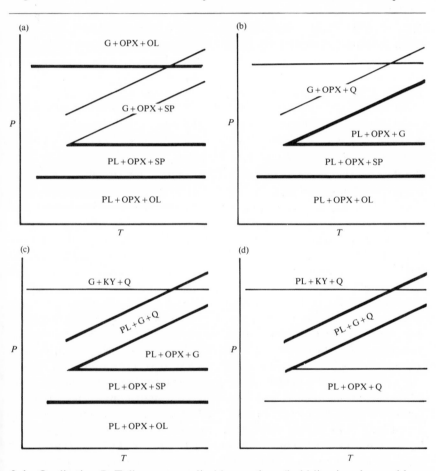

9.4 Qualitative *P–T* diagrams, applicable reactions (bold lines) and assemblages for different composition igneous rocks; see WE 9(a).

involved in constructing the diagrams. G–CPX rocks occur over a much wider range of compositions than is implied by the diagrams, mainly because of the range of solid solution observed in both garnet and clinopyroxene.

9(b) Discuss the depth of origin of the main basaltic magma types in the mantle in terms of a_{SiO_2} in the magma.

In WE 5(f), a $P-T$ diagram was contoured for the activity of silica in a magma in equilibrium with olivine and orthopyroxene in the mantle at that temperature and pressure (figure 5.8(a)). Figure 5.8(b) is an equivalent diagram in which the way a_{SiO_2} changes with pressure in the magma is taken into account so that the a_{SiO_2} in a magma at the Earth's surface (calculated from the mineralogy) can be compared directly with the contours.

First we will consider magmas generated by partial melting of mantle which then pass directly to the Earth's surface without fractionating (i.e. without changing composition). In this case, the a_{SiO_2} of the magma when it reaches the Earth's surface can be compared directly with the appropriate contour on figure 5.8(b) to give the pressure and temperature at which the magma could have been in equilibrium with the mantle if the ascent path of the magma is known. The ascent path is unlikely to be outside the 0·3 to 2 °C/km range and so the depth is known within a range. There are several points to be noted here.

The first is that as a piece of mantle melts, the composition of the remaining mantle minerals changes. Fortunately the a_{SiO_2} contours in figure 5.8(b) are relatively insensitive to this because the olivine and orthopyroxene change composition sympathetically so that the ratio of the olivine and orthopyroxene activities changes little as partial melting proceeds. The second point concerns determining the activity of silica in the lava. The magma that forms a lava flow will usually contain some phenocrysts formed in a near-surface magma chamber or in the conduit on the way to the surface. On extrusion the lava cools rapidly, the result being the groundmass mineralogy. The activity of silica in the lava will change progressively as crystallization proceeds so that the groundmass mineralogy crystallized from a liquid which need not have the same a_{SiO_2} as the magma which arrived at the Earth's surface. For example, if olivine is on the liquidus of a basalt (i.e. olivine crystallizes first when the liquidus is intersected on cooling) and then is joined by orthopyroxene after the temperature has dropped a further 50 °C, the a_{SiO_2} of the liquid which is crystallizing both olivine and orthopyroxene will have been increased by the crystallization of olivine over that 50 °C interval. The a_{SiO_2} calculated from the orthopyroxene and olivine in the groundmass will be greater than the activity of silica in the original magma. This process is seen *par excellence* in layered basic intrusions where olivine and orthopyroxenes are cumulus phases (buffering ln a_{SiO_2} at approximately −0·4) while quartz appears as a late-stage intercumulus mineral, the presence of quartz buffering ln $a_{SiO_2} = 0$. Thus the activity of silica has increased quite dramatically during intercumulus crystallization. The problem of calculating the activity of silica in lavas becomes even more difficult if we try and use the felsic minerals (for example, nepheline + SiO_2 = albite, with ln a_{SiO_2} = −1·5) to fix a_{SiO_2} because felsic minerals usually start to crystallize well below the liquidus, so the liquid can have changed composition substantially by the time they crystallize. Some progress can be made if the sense of the change of activity of silica with crystallization can be obtained. For example, the crystallization of olivine increases activity of silica, but note that crystallization of both olivine and orthopyroxene will not change activity of silica as this assemblage buffers activity of silica.

Allowing for an increase of a_{SiO_2} during the appearance of sufficient phases from which a_{SiO_2} in the lava can be calculated (say 0·3 ln units), a temperature–depth diagram can be drawn on which certain contours can be labelled with assemblages which would develop in a lava at the Earth's surface if the magma formed at that depth. This is done in figure 9.5, for

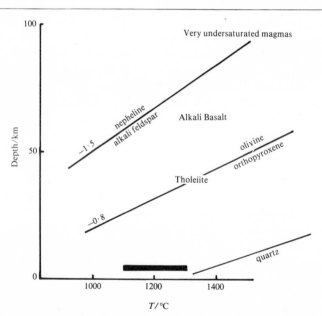

9.5   A depth–temperature diagram indicating the depths of origin of lavas which have particular assemblages when they crystallize at the Earth's surface, for certain simplifying assumptions; see WE 9(b).

the appearance of quartz, olivine + orthopyroxene, and nepheline + alkali feldspar. Tholeiites (and high alumina basalts) can contain olivine and orthopyroxene while alkali basalts rarely contain orthopyroxene, suggesting that alkali basalts come from deeper than tholeiites, if both formed in equilibrium with the mantle. More undersaturated rocks, which contain alkali feldspar + nepheline, or the really undersaturated rocks, containing, for example, leucite + kalsilite, or perovskite, can only be in equilibrium with the mantle at deeper levels. Quartz-bearing lavas cannot have formed in equilibrium with the mantle, given a certain thickness crust and the usual liquidus temperatures for lavas (bar on figure 9.5).

The above logic excludes any fractionation of the magma on the way to the Earth's surface. Low pressure (crustal) fractionation of most basaltic compositions leads to the formation of oversaturated rocks; for example basalt–andesite–dacite–rhyolite, thus low pressure fractionation leads to an increase of a_{SiO_2}. This is one way of generating quartz-bearing lavas. For more undersaturated alkali basalts, low pressure fractionation leads to phonolites, with a decrease in a_{SiO_2}. The loss of fluid by a wet ascending magma is another form of fractionation which may affect a_{SiO_2}. One suggestion is that oversaturated lavas may be generated by wet melting of mantle. However it is not clear whether there is ever sufficient water in the mantle for this purpose, or whether water can be lost from the ascending magma in the prescribed way.

High pressure fractionation may take several forms. If the magma does not equilibrate with the mantle but crystallizes in isolation from it (closed system fractionation), then the result will depend on the crystallizing phases. For example, basaltic compositions will fractionate eclogite around 25 kbar, which has the effect of decreasing a_{SiO_2} in the magma. On the other hand, if the magma equilibrates with the mantle during ascent (open system fractionation), then the magma will attain an a_{SiO_2} value for the mantle appropriate to these conditions. Any previous a_{SiO_2} history is lost. Thus, a very undersaturated magma will take on a tholeiitic a_{SiO_2} value if it equilibrates with the mantle at 50 km and 1400 °C. The same logic applies to a magma which rises slowly through the mantle remaining in equilibrium with it. The a_{SiO_2} for a lava formed in this way corresponds to the temperature and pressure at which the magma separates from the mantle and moves more rapidly towards the Earth's surface. The depth of origin calculated using the a_{SiO_2} method thus refers to the depth at which the magma separated from the mantle, not the depth at which the first partial melt formed. This relates to one of the main processes of generating magmas from the mantle. If a volume of mantle rises (for example, as part of a convection cell) then partial melting is likely to start to occur because the reduction of pressure is much faster than cooling (because of thermal inertia). In this way, the solidus of the rock can be intersected. Once melting has started, the degree of partial melting increases as the rock continues to rise until the magma can segregate and move towards the Earth's surface more rapidly. The a_{SiO_2} method will give information on the conditions when segregation occurs.

The activities of the other major oxides will operate in the same way as SiO_2, tending not to reflect any previous history. Thus an undersaturated magma formed at great depth, continuously equilibrating with the mantle to 50 km will look like a tholeiite from a major element point of view. However this will not be the case for trace elements, particularly those which prefer to be in the magma, the incompatible elements. The content of such trace elements in the magma will be the sum of all the processes which have affected the magma, rather than just the result of the last process. The magma effectively leaches the incompatibles out of the mantle with which it equilibrates.

Note that different degrees of partial melting of mantle at a particular temperature and pressure will produce magmas with more or less the same activity of silica because this is buffered by the olivine and orthopyroxene in the mantle. Nevertheless this does not mean that the amount of SiO_2 (or the other oxides) in the liquids will be the same. Similarly, just because the mantle can generate a magma with an activity of silica appropriate to a tholeiite does not mean that that magma will be a tholeiite. This will depend on the amounts of all the other oxides. At least experimental petrology has shown us that basaltic magmas are produced by partial melting of possible mantle materials at the appropriate temperatures and pressures.

The main basaltic magma types, tholeiitic to alkali basaltic, last equilibrated with the mantle at depths shallower than 75 km, although the processes forming them may have been initiated at greater depths than this.

9(c) Use the $T–x$ phase diagrams in figure 9.6 to model features of melting and crystallization of magmas generated by partial melting from an olivine–orthopyroxene source.

The two phase diagrams are for $Mg_2SiO_4–SiO_2$ at low and high pressure, using OPX for $MgSiO_3$, OL for Mg_2SiO_4 and Q for SiO_2. In the Mg system, the change of topology, produced by R moving across the OPX composition to become F, occurs at 2 or 3 kbar. As other end-members are added, this pressure increases, and we shall assume that processes in the upper mantle can be modelled using figure 9.6(a) (called high pressure), and that crustal processes can be modelled using figure 9.6(b) (called low pressure). The crucial thing about these diagrams is that the OPX composition is a plane of silica saturation, lavas to the left of OPX crystallizing to Q-bearing assemblages, to the right of OPX to OL-bearing assemblages.

An approximate mantle composition is marked M. At high pressures, heating M results in the generation of a first melt at F at T_1 by the reaction $OL + OPX = L_F$. With further melting the liquid becomes more magnesian until a melt of composition M is generated. The important thing is that all melt compositions produced at high pressure are silica-undersaturated (i.e. to the right of OPX). Further, the equilibrium or fractional crystallization of any of these liquids at high pressures will always involve the disappearance of the liquid by the reaction $L_F = OL + OPX$, so that there is no way of producing silica-oversaturated liquids at high pressure in this system. (This does not mean that, for example, the addition of water will not complicate matters and, under special circumstances, allow silica-saturated magmas to be generated in the mantle.)

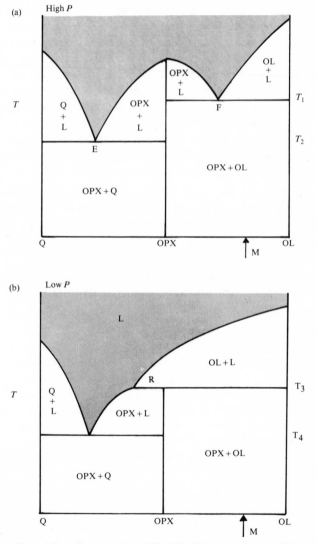

9.6   Schematic melting diagrams for SiO_2–Mg_2SiO_4 at low and high pressure. Q used for all SiO_2 polymorphs, and EN for all $MgSiO_3$ polymorphs. (WE 9(c))

At low pressure, R has a special significance with respect to crystallization of magmas. Consider a magma formed at high pressure of composition F, crystallizing at low pressure. For equilibrium crystallization, olivine will crystallize from the liquid down to T_3, the liquid changing composition to R. At T_3, a reaction takes place, $L_R + OL = OPX$, and if equilibrium is maintained, the liquid L_R is used up at this temperature, leaving the assemblage $OL + OPX$. The proportions of OL and OPX in the final assemblage can be determined with the lever rule using the composition F. For fractional crystallization, the OL generated in the liquid F down to T_3 is continuously removed, so that at R, the equilibrating system has the

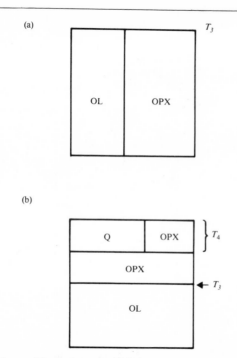

9.7 Imaginary layered intrusions produced by crystallizing liquids A and B and settling the resulting crystals, at low and high pressure. (WE 9(c))

composition of R. Thus OPX starts to crystallize at R and the liquid changes composition to E, until at T_4, Q enters the assemblage by the reaction $Q + OPX = L_E$. All the liquid is used up at T_4 and the final assemblage is $Q + OPX$, the proportions being given by the lever rule for composition E. The equilibrium and fractional crystallization of F at low pressure can be summarized by imaginary intrusions of initial composition F; figure 9.7(a) being for equilibrium crystallization, figure 9.7(b) being a layered intrusion generated by fractional crystallization and sinking of the minerals as they are produced.

The important thing is that low pressure fractional crystallization of mantle-generated magmas, for example F, can produce silica-oversaturated magmas by fractional crystallization. The sequence of lavas basalt–andesite–dacite–rhyolite found in many basaltic volcanoes is then best explained in terms of generation of the basaltic magma in the upper mantle, followed by fractional crystallization in a crustal magma chamber.

9(d)   Predict the phase relations in pyroxenes for slow subsolidus cooling in rocks along the differentiation trend in figure 9.8.

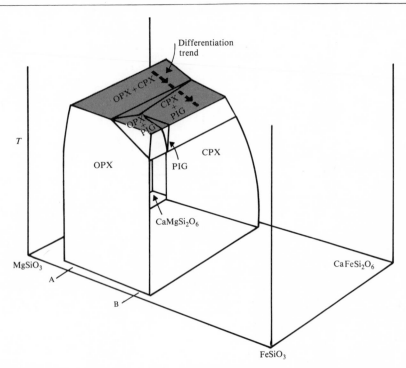

9.8 Schematic block diagram of pyroxene subsolidus relations in $MgSiO_3$–$FeSiO_3$–$CaMgSi_2O_6$–$CaFeSi_2O_6$ applicable to Skaergaard pyroxene textures. The solidus is represented by the shaded plane. (WE 9(d))

The differentiation trend and the sequence of pyroxene assemblages at the solidus correspond to those found in Skaergaard and other basic layered intrusions. The more magnesian, early formed rocks contain an orthopyroxene (OPX) and Ca-rich clinopyroxene (CPX); while the later formed, more Fe-rich rocks contain a pigeonite (PIG) and a Ca-rich clinopyroxene (CPX), both at the solidus. At the transition between the assemblages should be the assemblage OPX–CPX–PIG.

The likely subsolidus phase relations can be best identified from T–x

diagrams for appropriate stages in the differentiation trend. Although strictly these diagrams are pseudobinary, the phases involved are not far removed from the plane of the $T-x$ diagram so that the diagram can be treated as a section, with information on compositions and proportions obtainable from the diagram.

In figure 9.9(a), OPX + CPX are the solidus pyroxenes, their compositions being given by the limbs of the miscibility gap at each temperature. As

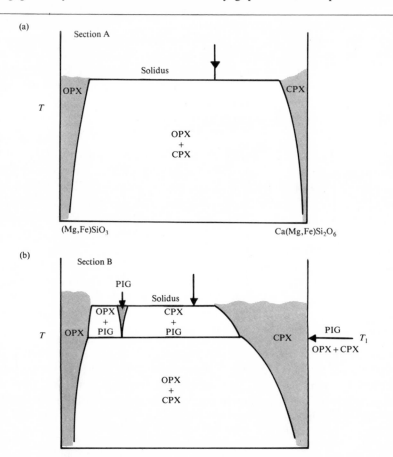

9.9 $T-x$ sections through the block diagram, figure 9.8, at A and B, (a) and (b) respectively, used to predict pyroxene relations with cooling. (WE 9(d))

each of the minerals cools, their equilibrium compositions become progressively more extreme. This is attained by exsolution, the OPX exsolving lamellae of CPX, the CPX exsolving lamellae of OPX. The minerals will continue to exsolve with falling temperature until below a closing temperature, the minerals no longer change composition and equilibrium will not be attained. The closure temperature will depend on the ease of diffusion of ions in the pyroxenes and the cooling rate of the rock. For fast subsolidus cooling, as in a lava, no exsolution will take place, at least on a macroscopic scale.

In figure 9.9(b), CPX + PIG are the solidus pyroxenes, their compositions being given by the limbs of the miscibility gap at each temperature. During cooling to T_1, the PIG exsolves CPX and the CPX exsolves PIG. At T_1, the reaction PIG = OPX + CPX takes place resulting in the breakdown of PIG to OPX and CPX, in approximate proportion 4OPX:CPX. The form of the breakdown products is controlled by the fact that the OPX structure is not very different from the PIG structure, so that each PIG grain is replaced by an OPX grain with abundant CPX lamellae. This process is usually referred to as pigeonite inversion to orthopyroxene, although the term inversion is usually used when the reaction is polymorphic, no composition change being involved (aragonite inverting to calcite, and so on).

Below T_1, the OPX exsolves CPX and the CPX exsolves OPX until the closure temperature is passed. The resulting CPX grains will have high temperature PIG lamellae which inverted to OPX + CPX at T_1 and later OPX lamellae. The resulting OPX grains will have early CPX lamellae exsolved from the original PIG and CPX lamellae which were produced by the PIG breakdown, usually extended by further exsolution to lower temperature. The early CPX lamellae are often in a different crystallographic orientation to the later lamellae. These relations can be seen in thin sections of, for example, Skaergaard MZ gabbros. On a submicroscopic scale, electron microscopy of pyroxenes has shown a bewildering variety of lamellae representing exsolution during the different stages of cooling.

In figure 9.8, the appearance of PIG on the solidus is the result of the slope on the PIG = OPX + CPX reaction, moving to lower temperatures in more Fe-rich systems. This reaction has moved down from igneous to upper amphibolite facies metamorphic conditions in more Fe-rich systems so that pigeonites and 'inverted' pigeonites are found in metamorphic iron formations.

9(e) The Barrovian zone sequence of assemblages for pelitic rocks is summarized in figure 9.11 using the AFM projection (figure 9.10(a)). Show how the metamorphic assemblage changes with increasing temperature for different rock compositions and identify useful isograds.

The AFM projection gives a compatibility diagram for showing assemblages in rocks containing quartz (Q), muscovite (MU) and, originally, an H_2O fluid phase. Muscovite is often a solid solution formed by substituting AlAl = MgSi in $KAl_3Si_3O_{10}(OH)_2$; these micas are called phengites.

This complicating factor in the projection is ignored in figure 9.10(a), and does not seriously affect the logic presented here.

The first step in considering the way the assemblages change with temperature is to see what are the usual compositions of pelitic rocks

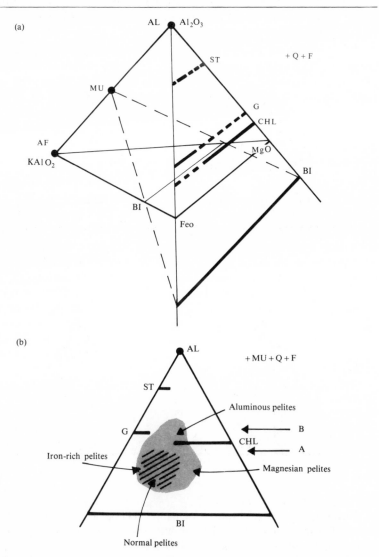

9.10 The quartz, fluid-projected $KAlO_2$–Al_2O_3–FeO–MgO compatibility tetrahedron, (a), for pelitic rocks, showing the positions of the phases, and the projection from MU onto the MgO–FeO–Al_2O_3 face of the tetrahedron. (b) is this $MU + Q + H_2O$ projection showing the composition range which includes most pelitic rocks. (WE 9(e))

(figure 9.10(b)). As in WE 9(a), the sequence of assemblages developed with changing conditions will be strongly dependent on rock composition. Turning to figure 9.11, there are several points to note. First, the compositions of the phases in particular assemblages are changing with temperature, for example, the biotite (BI) and chlorite (CHL) compositions in the AL–CHL–BI assemblage from (c) to (d) (AL stands for

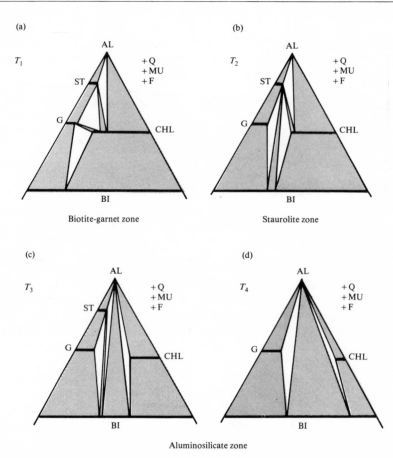

9.11 Four $MU + Q + H_2O$ projections at equal temperature intervals representing Barrovian-type metamorphism of pelites. Garnet–biotite–feldspar equilibria for very Fe-rich compositions, the stability of chloritoid at low temperature and the stability of cordierite are ignored. (WE 9(e))

aluminosilicate, ST for staurolite and G for garnet). Second, the disposition of tie lines changes between each succeeding diagram. From (a) to (b), the reaction is:

$$G + CHL = ST + BI$$

From (b) to (c) the reaction is:

ST + CHL = AL + BI

From (c) to (d) the reaction is:

ST + BI = G + AL   then   ST = G + AL

or

ST = G + AL   then   ST = G + AL + BI

The sequence of assemblages with changing temperature for each rock composition is controlled by these continuous and discontinuous changes in the positions of tie lines with temperature.

To see these effects properly, we need to construct $T–x$ diagrams to summarize the information in figure 9.11, and to interpolate between these diagrams. There are several ways of doing this. The first is a pseudobinary $T–x$ diagram involving the assemblages developed for a particular section through AFM with changing temperature; in this case for the horizontal sections at A and B in figure 9.10(b). The section is only *pseudo*binary because the compositions of the phases in the assemblage do not plot in the plane of the section. The resulting $T–x$ diagram will give information on the assemblages but not on the compositions of the phases in each assemblage.

Figure 9.12 is the $T–x$ diagram for section A, thus including the assemblages for 'normal' pelites. Clearly there is a certain amount of filling in required between the four compatibility diagrams given in figure 9.11, but the form of the diagram is certainly consistent with figure 9.11. The boundaries of the three-phase fields slope because of the continuous change in the positions of tie lines with temperature. The horizontal lines across which the three-phase assemblages change, reflect the occurrence of the reactions mentioned above. Thus at T_A, the reaction G + CHL = ST + BI means that rock compositions in the section between $x_{Mg} = 0.15$ and 0.3 go from G + CHL + BI to G + ST + BI while compositions between $x_{Mg} = 0.3$ and 0.45 go from G + CHL + BI to CHL + ST + BI, both with increasing temperature. Assemblages more magnesian or less magnesian than the range $x_{Mg} = 0.15$ to 0.45 will not be affected by the reaction. The sequence of assemblages for each rock composition along the section A can be obtained by reading off the assemblages along a vertical line at the appropriate composition in figure 9.12.

For a 'normal' pelite, with say $x_{Mg} = 0.3$ (marked by an arrow in figure 9.12), the sequence with increasing temperature is CHL + BI, G + CHL + BI, G + ST + BI, AL + G + BI. The appearance of garnet in pelites is often taken as an isograd (the garnet isograd), but we can see here that the appearance of garnet is dependent on the composition of the pelite, the more Fe-rich the rock, the earlier the appearance of garnet. The appearance of garnet is therefore a most unsatisfactory isograd. (The appearance of garnet is also affected by the amount of Mn available, as Mn-bearing garnet becomes stable at a lower temperature than Mn-free garnet for a particular rock composition.) On the other hand, the appearance of

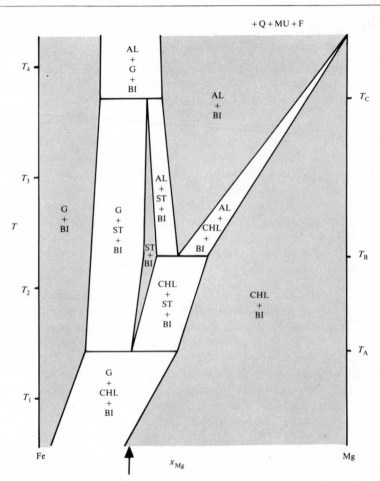

9.12 $T-x$ pseudobinary section for A in the compatibility diagram, figure 9.10(b). The arrow marks a composition discussed in the text. (WE 9(e))

staurolite by the reaction $G + CHL = ST + BI$ is a good isograd because it occurs at a particular temperature (for a particular pressure) regardless of rock composition, at least in the AFM system. However, the appearance of staurolite in more magnesian pelites ($x_{Mg} > 0.45$) will depend on rock composition. The appearance of an aluminosilicate in an assemblage by the reaction $ST = G + AL + BI$ is a good isograd for the same reason. Note that for more magnesian pelites, an aluminosilicate will appear in the assemblage at T_B rather than T_C by the reaction $ST + CHL = AL + BI$, again making a good isograd.

Summarizing the above, good isograds are formed by discontinuous changes of the positions of tie lines, not by continuous changes of the positions of tie lines. To identify the position of such an isograd, it is necessary to observe the appropriate tie lines on one side of the isograd changing to the new tie lines on the other side. It is not sufficient just to observe the appearance of a new mineral. A good isograd can only be really useful if it occurs in common rock compositions, like the above staurolite isograd.

These conclusions can be emphasized using the equivalent $T-x$ diagram for section B in figure 9.10(b), applicable to aluminous pelites (figure 9.13). It is constructed in the same way as figure 9.12. The only ambiguity is in the reactions involved in the disappearance of staurolite in very Fe-rich compositions. The $G + AL + ST$ loop occurs at a lower temperature than T_C if the reaction removing staurolite is $ST = G + AL + BI$, whereas it will occur at a higher temperature than T_C if the reaction is $ST = G + AL$. Comparing figures 9.12 and 9.13, the assemblages at lower temperature are quite different, nevertheless the reactions still occur at the same temperature in each. Note that in aluminous systems, an aluminosilicate will appear in the assemblage at a much lower temperature than in normal pelites. The appearance of an aluminosilicate in a pelite certainly does not form an isograd unless it appears by a discontinuous change in the position of tie lines.

Another sort of diagram which allows much more information on mineral compositions in the assemblages as a function of temperature is the $T-x$ projection. As all the minerals of interest with variable composition can be treated as just Fe–Mg solid solutions, these compositions may be plotted as the x-axis of a $T-x$ diagram. Composition information for all the phases in all the three-phase assemblages in the AFM diagram can now be plotted on the same diagram (figure 9.14). It now becomes more difficult to locate the lines for minerals which belong to the same assemblage, but the amount of information contained on the diagram is much larger. The lines for a particular assemblage terminate on the same reaction line. Thus, for the $G + CHL = ST + BI$ reaction, there are two sets of lines coming in from low temperature (at different angles), one set each for the assemblages G–CHL–BI and G–CHL–ST; and two sets of lines leaving at high temperature, one set for G–ST–BI which goes up to the reaction at T_C, the other for CHL–ST–BI which goes up to the reaction at T_B. Those sets involving only two lines for an assemblage contain an aluminosilicate as the third phase, which cannot be projected onto this diagram. This type of diagram provides an excellent way of interpolating between compatibility diagrams in a system involving phases of variable composition.

9(f) Consider high-grade metamorphism and melting of pelites using figure 9.15(a), representing the pelites by the simplified system $KAlO_2-Al_2O_3-SiO_2-H_2O$ involving assemblages with alkali feldspar (AF), muscovite (MU), sillimanite (S), quartz (Q), H_2O fluid phase (F) and silicate liquid (L).

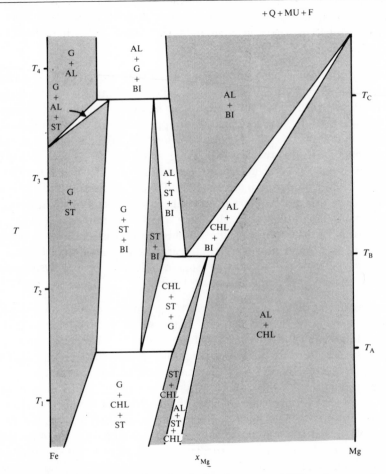

$+Q+MU+F$

9.13 $T-x$ pseudobinary section for B in the compatibility diagram, figure 9.10(b). (WE 9(e))

This simple system can be used to model behaviour in the Na–K system, the pressure and temperature for the intersection X corresponding to conditions for a 'natural' Na–K system rather than the pure K system, which occurs at higher temperature. The simplest compatibility diagram for this system is figure 9.15(b), a projection from Q. These compatibility diagrams can then be inserted on the *P–T* diagram. The liquid is represented by a single composition, whereas, with increasing temperature, the area of possible liquid compositions increases until the area extends over the whole diagram at some higher temperature (at the liquidus). Pelites of different compositions melt at different temperatures. The line of beginning of melting (the solidus) for the Q–MU–S–F assemblage is

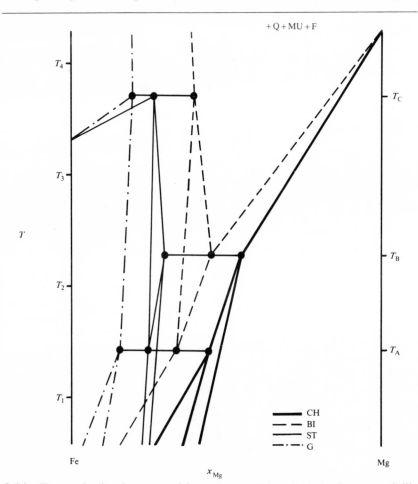

9.14 *T–x* projection for compositions between BI and AL in the compatibility diagram, figure 9.10(b), showing the compositions of the phases in each three-phase field. (WE 9(e))

[AF]–[MU], for the Q–MU–AF–F assemblage is [S]–[MU], and for the fluid-absent assemblage, Q–S–AF–MU, is [F]–[MU].

A and B on figure 9.15(b) can be used to represent the two fluid-present

assemblages, with the proportion of fluid in the assemblages being small. For an increase of temperature at P_1, the amount of melting of A when it crosses [AF] will be small because A plots close to the MU–S side of the MU–S–L triangle. The amount of melting increases dramatically when [F] is crossed because A now plots centrally within the AF–S–L triangle. Similarly for B, the amount of melting produced by crossing [S] is small,

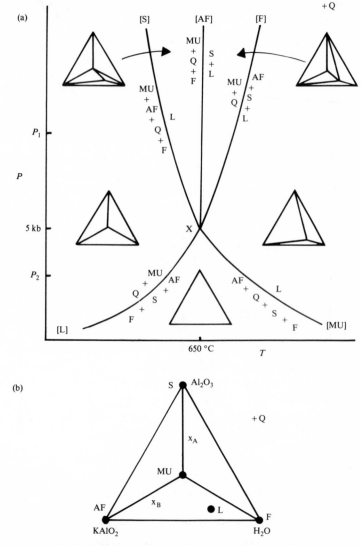

9.15 Schematic P–T diagram, (a), showing relations in $KAlO_2$–Al_2O_3–H_2O involving MU, AF, Q, F and L using the compatibility diagram, (b). Compositions A and B are discussed in the text. (WE 9(f))

but increases substantially when [F] is crossed. Thus the line along which substantial melting occurs is [F]–[MU] for fluid-present (as well as fluid-absent) rocks. The small amount of melting produced before [F] is crossed (above 5 kbar) may have helped give rise to some of the banded structures and pegmatites in high-grade pelites, before the more substantial amount of melting associated with [F] gives rise to migmatites and then to granites, if the silicate melt can separate from the residue (or carry at least some of the residue along with it) and rise higher in the crust.

The residue from crossing [F] or [MU] and removing the liquid will be $Q+S+AF$, a fluid-absent assemblage which is in fact a granulite facies pelitic assemblage. Thus a pelite in the amphibolite facies (Q–AF–MU–F) can make the prograde transition into the granulite facies (Q–AF–S) if the silicate liquid formed at [F] or [MU] is removed. A similar process is needed to get any hydrous assemblage into the granulite facies because of the relatively low temperature of the appearance of silicate liquid in fluid-present systems.

Fluid inclusions are found in granulite facies pelites, but they are CO_2-rich. This can be accounted for because silicate liquids at crustal pressures have much lower solubilities of CO_2 than H_2O. Figure 9.16(a) shows the stability relations at a lower temperature than [S] and [L] in figure 9.15(a). The fluid composition a can be in equilibrium with $MU+Q+AF+S$ and so this corresponds to conditions appropriate to [L] for a fluid of composition a. The assemblage in equilibrium with a fluid more CO_2-rich than a is $Q+AF+S$. Figure 9.16(b) shows the stability relations at a higher temperature than [F] and [MU] in figure 9.15(a). $L+AF+S+Q$ can be in equilibrium with a H_2O–CO_2 fluid of composition b. Removal of the liquid from this assemblage will leave $AF+Q+S+F$, where F has composition b. Thus, the presence of CO_2-rich fluid inclusions in granulite facies pelites is quite consistent with an origin by prograde metamorphism of a fluid-present pelite, with subsequent loss of the partial melt formed in crossing [F] or [MU].

9(g) Discuss the retrograde metamorphism of the assemblage, sillimanite (AL), alkali feldspar (AF), quartz (Q) and fluid (F).

There are two main aspects to the retrograde metamorphism of this assemblage. The first concerns the retrogression of sillimanite to either kyanite or andalusite (depending on the pressure). Looking at figure 7.8, these reactions are overstepped by a considerable extent, so that it is possible that nothing happens to the sillimanite from these reactions, particularly if the rate of nucleation and growth decrease to lower temperatures as expected.

The second aspect concerns the hydration reaction $AF+AL+H_2O = MU+Q$, the muscovite+quartz breakdown reaction approached from the high temperature side. Retrogression will involve the appearance of muscovite in the assemblage. The simple-minded way of looking at how much muscovite is developed during retrogression is to consider how much H_2O is available to perform the necessary hydration. For example if there is 1 mole % H_2O in the assemblage when the reaction is intersected

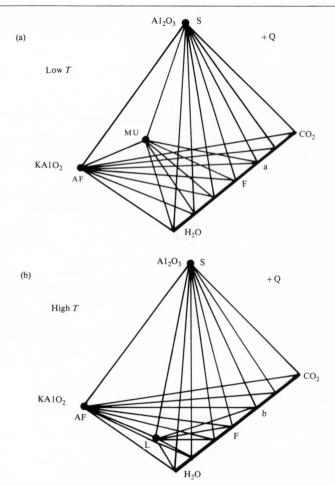

9.16 Tetrahedral compatibility diagrams showing distribution of tie lines at low and high temperature, (a) and (b) respectively. (WE 9(f))

as temperature decreases, then the maximum amount of muscovite that can be developed is about 1 mole %. If the terminal stages of prograde metamorphism involve a general increase of grain size (associated with the recrystallization of strained grains?), then the proportion of H_2O may well be less than this, and so the maximum amount of muscovite would also be less.

The actual amount of muscovite developed may be less than the maximum amount because the fluid phase composition is likely to be buffered by the assemblage during retrograde metamorphism. Figure 9.17 is the 4 kbar $T-x$ diagram obtained from the contours on figure 6.5. The internal buffering equation can be used to determine the amount of muscovite developed through a particular temperature interval (i.e. through a particular x_{H_2O} interval).

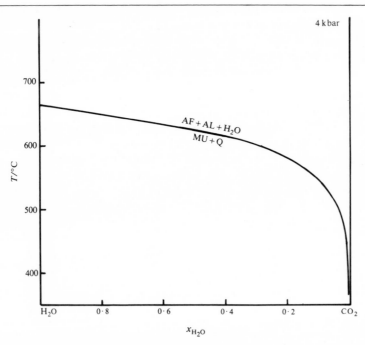

9.17 $T-x_{CO_2}$ diagram at 4 kbar for the $MU+Q$ breakdown reaction used to illustrate retrograde metamorphic effects in WE 9(h).

The equation to use is the exact equivalent of the CO_2 one, i.e.:

$$\Delta n_{H_2O} = \frac{a}{1 - Ax_{H_2O}} \Delta x_{H_2O}$$

where $A = (y_{CO_2} + y_{H_2O})/y_{H_2O}$; in this case $A = 1$. If we take $a = 1$ (to represent the likely maximum amount of fluid), and noting that $\Delta n_{MU} = -\Delta n_{H_2O}$:

$$\Delta n_{MU} = -\frac{1}{1 - x_{H_2O}} \Delta x_{H_2O}$$

The maximum on this reaction at $x_{H_2O} = 1$ affects retrograde metamorphism in the reverse of its effect during prograde metamorphism. If the assemblage intersects the reaction near the maximum during cooling, then a substantial amount of MU is developed very quickly. Thus taking $x^\circ_{H_2O} = 0.95$, then in 10 °C ($x_{CO_2} = 0.8$), $\Delta n_{MU} = 0.75$, i.e. about 80% of the possible muscovite that can form. By $x_{CO_2} = 0.5$, 95% of the possible muscovite has formed. However if the fluid is substantially more CO_2-rich, say with $x^\circ_{H_2O} = 0.6$, then for a 100 °C drop in temperature ($x_{CO_2} = 0.15$) only just over 50% of the possible muscovite has formed.

If the reaction being followed intersects other reactions, then the path of the assemblage can be followed down temperature in the same way as for prograde metamorphism in WE 8(a). The above discussion has assumed equilibrium is obtained at each stage. At some stage during cooling, the rate of equilibration will be overtaken by the rate of cooling so that equilibrium is no longer attained.

The points to note are that retrograde reactions which involve hydration or carbonation can only take place to a very limited extent if the fluid is supplied by the rock, and that the amount of hydration and/or carbonation need not be the maximum, depending on the initial fluid composition and the disposition of the retrograde reactions on a $T-x$ diagram. Clearly, if the assemblage of interest was infiltrated by a fluid, then the retrograde effects will be much more severe.

References

Ringwood, A. E., 1975, *Composition and Petrology of the Earth's Mantle*, McGraw-Hill, New York: Further reading on the mineralogy and petrology of basic and ultrabasic rocks from an experimental viewpoint (WE 9(a)).

Carmichael, I. S. E., Turner, F. J. and Verhoogen, J., 1974, *Igneous Petrology*, McGraw-Hill, New York: Further reading on the use of a_{SiO_2} in igneous petrology.

Wager, L. M. and Brown, G. M., 1968, *Layered Igneous Rocks*, Oliver and Boyd, Edinburgh. pp. 38–46: Source for compositional and textural information on coexisting pyroxenes in Skaergaard (WE 9(d)).

Thompson, A. B., 1976, Mineral reactions in pelitic rocks. *Am. J. Sci.*, **276**, 401–24: Source and advanced reading on the progressive metamorphism of pelites (WE 9(e)).

Luth, W. C., 1976, Granitic rocks. In D. K. Bailey and R. Macdonald (Eds), *The Evolution of the Crystalline Rocks*, Academic Press, New York: Source for WE 9(g).

Appendix A
Thermodynamic Properties

Introduction

The Gibbs energy of a phase A, G_A, is a measurable property of the phase which is dependent on temperature and pressure, the composition of A, and the crystal structure of A. The dependences on the composition and crystal structure are not specifically considered here, so the phase is considered to remain in the same structure and not change composition as we change temperature and pressure. The Gibbs energy is related to temperature and pressure by the enthalpy, H, the entropy, S, and the volume, V. The Gibbs energy at pressure P and temperature T, $G(P, T)$, is given by:

$$G(P, T) = H(1, T) - TS(1, T) + \int_{1}^{P} V(P, T) \, dP \qquad (A.1)$$

where $H(1, T)$ and $S(1, T)$ are the enthalpy and entropy at 1 bar and T, with P in bars and T in K. The first two terms give the Gibbs energy at 1 bar and T, the third takes this value up pressure. All the properties that will be referred to are molar properties. The units for Gibbs energy and enthalpy are kJ, for entropy are $kJ \, K^{-1}$, and for volume are $kJ \, bar^{-1}$. Two useful conversion factors are $1 \, kcal = 4 \cdot 184 \, kJ$ and $1 \, cm^3 = 0 \cdot 0001 \, kJ \, bar^{-1}$. Thermodynamic properties are often tabulated at 1 bar and 298 K (or 25 °C) so these conditions often appear in the various expressions in the following pages.

Volume

Volume is usually expressed as a function of P and T with the help of the coefficient of thermal expansion, α, and the isothermal compressibility, κ, which are defined by:

$$\alpha = \frac{1}{V}\left(\frac{\partial V}{\partial T}\right)_P \quad \text{and} \quad \kappa = -\frac{1}{V}\left(\frac{\partial V}{\partial P}\right)_T \tag{A.2a,b}$$

or, in words, α is the fractional change in volume, dV/V, with respect to changing temperature at constant pressure, while κ is the fractional change in volume with respect to changing pressure at constant temperature. For solids, α and κ change slowly with increasing temperature and pressure. For simplicity, we will assume that they are independent of temperature and pressure and we will integrate (A.2a,b) in turn. Integrating (A.2a) at 1 bar:

$$\int_{298}^{T} \alpha \, dT = \int_{V(T=298)}^{V(T=T)} \frac{1}{V} \, dV$$

gives:

$$\alpha(T-298) = \ln\left[\frac{V(1, T)}{V(1, 298)}\right]$$

Re-arranging:

$$V(1, T) = V(1, 298) \exp\left[\alpha(T+298)\right] \tag{A.3}$$

Integrating (A.2b) at constant temperature:

$$-\int_{1}^{P} \kappa \, dP = \int_{V(P=1)}^{V(P=P)} \frac{1}{V} \, dV$$

gives:

$$-\kappa(P-1) = \ln\left[\frac{V(P, T)}{V(1, T)}\right]$$

Re-arranging, and noting that as P is usually much larger than 1, so that $P-1$ can be replaced by P, then:

$$V(P, T) = V(1, T) \exp\left(-\kappa P\right)$$

Substituting (A.3)

$$V(P, T) = V(1, 298) \exp\left[\alpha(T-298)\right] \exp\left(-\kappa P\right)$$
$$= V(1, 298) \exp\left[\alpha(T-298) - \kappa P\right]$$

As the term in square brackets is very small, then, using $\exp(x) = 1 + x$,

$$V(P, T) = V(1, 298)[1 + \alpha(T-298) - \kappa P] \tag{A.4}$$

The units of α are K^{-1}, of κ are bar^{-1}.

The volume term in ΔG° is the integral of (A.4) with respect to pressure. Thus, again assuming that $P-1$ can be replaced by P, then:

$$\int_1^P V(P, T)\,dP = \int_1^P V(1, 298)[1 + \alpha(T-298) - \kappa P]\,dP$$

$$= PV(1, 298)[1 + \alpha(T-298) - \kappa P/2] \qquad\qquad (A.5)$$

This can be compared with the volume integral if the volume is assumed to be independent of temperature and pressure. Then:

$$\int_1^P V(P, T)\,dP = \int_1^P V(1, 298)\,dP = PV(1, 298) \qquad\qquad (A.6)$$

For example consider the volume integral for Mg_2SiO_4, forsterite,

$V(1, 298) = 0.004379 \text{ kJ bar}^{-1}$ (Robie $et\ al.$ 1967)

$\alpha = 41 \times 10^{-6} \text{ K}^{-1}$

$\kappa = 0.8 \times 10^{-6} \text{ bar}^{-1}$

(assuming α and κ are independent of P–T, using tables in Clark 1966). The volume integral (A.5) at 5000 bar and 1073 K is:

$0.004379 \times 5000[1 + 41 \times 10^{-6}(1073 - 298) - 0.8 \times 10^{-6} \times 5000/2]$
$$= 22.55 \text{ kJ}$$

If the volume is assumed to be independent of pressure and temperature, then the volume integral (A.6) for this example is:

$0.004379 \times 5000 = 21.90 \text{ kJ}$

As temperature and pressure are increased, the difference between (A.5) and (A.6) increases, as can be seen in figure A.1. The error involved in using (A.6) rather than (A.5) at crustal temperatures and pressures is insignificant for most purposes.

For fluids, the volume is a strong function of temperature and pressure such that α and κ are both functions of temperature and pressure. The above approach is no longer useful. Instead the volume integral is tabulated at a range of temperatures and pressures.

Entropy

Entropy is a property which reflects the degree of disorder in a system; the more disordered the system, the higher its entropy. Disorder here refers to positional disorder, whether the atoms are free to move as in a gas or have fixed positions on a lattice as in a crystal; vibrational disorder, whether the atoms vibrate about an average position usually due to thermal motion; and so on. From the first type of disorder we can see that for a given composition, a mole of gas will have a larger entropy than a mole of liquid (which can be considered to have a quasi- or imperfect lattice) which, in turn, will have a larger entropy than a mole of solid. For example for H_2O, $S(1, 298)_{gas} = 0{\cdot}1887$ kJ K^{-1}, $S(1, 298)_{water} = 0{\cdot}0699$ kJ K^{-1}, and $S(1, 298)_{ice} = 0{\cdot}0447$ kJ K^{-1}. From the second type of disorder we can see that the entropy of a solid (or liquid or gas) will increase with temperature because thermal motion increases with temperature. For solids, a particularly important type of disorder is configurational disorder. This refers to the distribution of different atoms on the

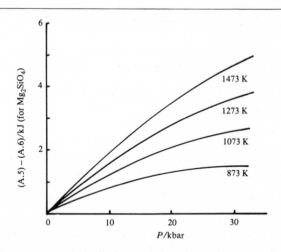

A.1 The difference between the volume integrals given by (A.5) and (A.6) plotted against pressure and contoured for temperature for Mg_2SiO_4, olivine, showing the error introduced by assuming that V is independent of P and T.

several different sites in the lattice. Consider the one tetrahedral and two octahedral sites in a AB_2O_4 spinel. The spinel structure can be ordered by (1) putting the A atom on the tetrahedral site, and the two B atoms on the octahedral sites (the 'normal' spinel distribution); or, (2) putting one B atom on the tetrahedral site, and mixing the remaining B atom and the A atom on the octahedral sites (the 'inverse' spinel distribution); or be disordered, with the A and B atoms distributed on both octahedral and tetrahedral sites (the 'disordered' spinel distribution). The entropy of a phase will depend on the distribution of the atoms on the sites, the more configurational disorder, the higher the entropy. The entropy of a particular spinel will increase from 'normal', through 'inverse' to 'disordered' depending on the actual site distributions.

The entropy is related to temperature by the heat capacity at constant pressure, C_p:

$$S(1, T) - S(1, 0) = \int_0^T \frac{C_p(1, T)}{T} \, dT$$

All forms of order are supposed to become complete as a phase approaches absolute zero. The third law of thermodynamics states that the entropy of a phase at absolute zero is zero, or $S(1, 0) = 0$, so that:

$$S(1, T) = \int_0^T \frac{C_p(1, T)}{T} \, dT$$

This equation is often considered in two parts:

$$S(1, T) = \int_0^{298} \frac{C_p(1, T)}{T} \, dT + \int_{298}^T \frac{C_p(1, T)}{T} \, dT$$

$$= S(1, 298) + \int_{298}^T \frac{C_p(1, T)}{T} \, dT \qquad \text{(A.7)}$$

Low temperature heat capacity measurements are used to find $S(1, 298)$, while high temperature heat capacities (for example Kelley 1961) are used to calculate the entropy increment from 298 K to the temperature of interest. High temperature heat capacities are usually presented algebraically using the equation $C_p = a + bT - c/T^2$ where a, b and c are constants, different for each phase. Using this expression:

$$S(1, T) = S(1, 298) + \int_{298}^T \frac{a + bT - c/T^2}{T} \, dT$$

$$= S(1, 298) + a \ln \left(\frac{T}{298}\right) + b(T - 298) + \frac{c}{2}\left(\frac{1}{T^2} - \frac{1}{298^2}\right)$$

For example, for Mg_2SiO_4, forsterite:

$S(1, 298) = 0 \cdot 0952 \text{ kJ K}^{-1}$ (Robie and Waldbaum 1968)

$a = 0 \cdot 1498 \text{ kJ}$

$b = 2 \cdot 74 \times 10^{-5} \text{ kJ K}^{-1}$

$c = 3565 \text{ kJ K}^2$ (Kelley 1961)

Then, for example,

$$S(1, 1073) = 0 \cdot 0952 + 0 \cdot 1498 \ln (1073/298) + 2 \cdot 74 \times 10^{-5}(1073 - 298)$$
$$+ 3565(1/1073^2 - 1/298^2)/2$$
$$= 0 \cdot 0952 + 0 \cdot 1919 + 0 \cdot 0212 - 0 \cdot 0185$$
$$= 0 \cdot 2898 \text{ kJ K}^{-1}$$

Figure A.2(b) shows the entropy of forsterite as a function of temperature.

Enthalpy

Enthalpy is the energy which is associated with heat. For example the heat evolved when a substance is dissolved in acid is an enthalpy of solution. The heat evolved when substances react is an enthalpy of reaction. Implicit in these two examples is the idea that we are considering the amount of enthalpy (or heat) involved in going from an initial state to a final state in some process. However we need to know, for example, the enthalpy of Mg_2SiO_4, forsterite. The simplest way of defining this is with respect to forming Mg_2SiO_4 from something else, for example, the oxides or the elements. This 'something else', the reference state, is usually taken to be the pure elements in their most stable forms at the pressure and temperature of interest and we refer to the enthalpy of formation from the elements, $\Delta_f H$. For Mg_2SiO_4, the value of $\Delta_f H$ is the heat evolved from the formation reaction at the pressure and temperature of interest:

$$2Mg + Si + 2O_2 = Mg_2SiO_4$$

The $\Delta_f H$ values for many oxides are known from direct calorimetry on the formation reaction, for example $\Delta_f H_{Al_2O_3,\text{corundum}}$ has been obtained by measuring the heat involved in burning Al in O_2 to form Al_2O_3, corundum. $\Delta_f H$ values for silicates have been obtained by first finding the enthalpy of formation from the oxides. For example, for Mg_2SiO_4, this involves finding the enthalpy of solution of MgO, SiO_2 and Mg_2SiO_4, and summing these values to give an enthalpy of formation of Mg_2SiO_4 from the oxides. This is then combined with the known values of $\Delta_f H$ for the oxides to give the $\Delta_f H$ value for Mg_2SiO_4. Thus:

$$\Delta_f H_{Mg_2SiO_4} = -\Delta H_{\text{soln}}(Mg_2SiO_4) + 2\Delta H_{\text{soln}}(MgO) + \Delta H_{\text{soln}}(SiO_2)$$
$$+ 2\Delta_f H_{MgO} + \Delta_f H_{SiO_2}$$

Charlu et al. (1975) have measured the enthalpies of solution of MgO, SiO_2 and Mg_2SiO_4 in $2PbO.B_2O_3$ melt at 1 bar and 970 K. These are:

$$\Delta H_{\text{soln}}(Mg_2SiO_4) = 67 \cdot 4 \text{ kJ}$$
$$\Delta H_{\text{soln}}(MgO) = 4 \cdot 9 \text{ kJ}$$
$$\Delta H_{\text{soln}}(SiO_2) = -5 \cdot 1 \text{ kJ}$$

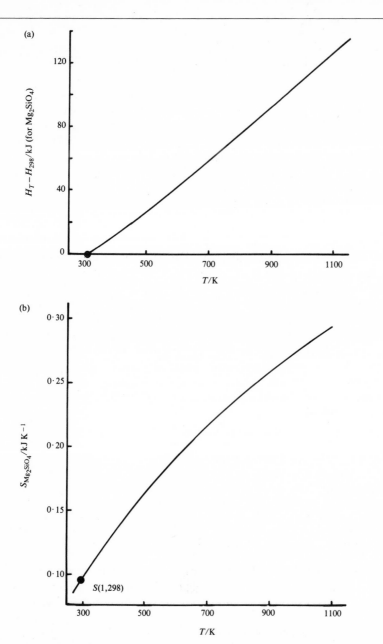

A.2 Plots of the heat content, $H_T - H_{298}$, and the entropy, S_T against temperature for Mg_2SiO_4, olivine, calculated using the equations in the text.

From Robie and Waldbaum (1968):

$$\Delta_f H_{MgO}(1, 970) = -609 \cdot 4 \text{ kJ}$$
$$\Delta_f H_{SiO_2}(1, 970) = -905 \cdot 1 \text{ kJ}$$

Thus:

$$\Delta_f H_{Mg_2SiO_4}(1, 970) = -67 \cdot 4 + 2(4 \cdot 9) + (-5 \cdot 1) + 2(-609 \cdot 4) + (-905 \cdot 1)$$
$$= -2186 \cdot 6 \text{ kJ}$$

The change of enthalpy of a phase with temperature, the heat content $H_T - H_{298}$, is calculated from the heat capacity using:

$$H_T - H_{298} = \int_{298}^{T} C_p(1, T) \, dT \qquad \text{(A.8)}$$

This gives the enthalpy change or heat content resulting from changing the temperature of the phase from 298 K to T K. Writing C_p in terms of a, b and c, (A.8) becomes:

$$H_T - H_{298} = \int_{298}^{T} (a + bT - c/T^2) \, dT$$
$$= a(T - 298) + b(T^2 - 298^2)/2 + c(1/T - 1/289)$$

For example, for Mg_2SiO_4, forsterite, using the heat capacity coefficients given previously:

$$H_{1073} - H_{298} = 0 \cdot 1498(1073 - 298) + 2 \cdot 74 \times 10^{-5}(1073^2 - 298^2)/2$$
$$+ 3565(1/1073 - 1/298)$$
$$= 116 \cdot 1 + 14 \cdot 6 - 8 \cdot 6$$
$$= 122 \cdot 1 \text{ kJ}$$

Figure A.2(a) shows the heat content of forsterite as a function of temperature.

The change of the enthalpy of formation with temperature must involve the heat capacity of the phase *and* its component elements. A heat capacity of formation is required, which, for forsterite, is given by:

$$\Delta_f C_{p,Mg_2SiO_4} = C_{p,Mg_2SiO_4} - 2C_{p,Mg} - C_{p,Si} - 2C_{p,O_2}$$

Then:

$$\Delta_f H(1, T) - \Delta_f H(1, 298) = \int_{298}^{T} \Delta_f C_p(1, T) \, dT = \Delta_f(H_T - H_{298})$$

For example, the enthalpy of formation value for forsterite at $(1, 970)$ calculated from calorimetric data above can be converted into a $(1, 298)$ value using:

$$(H_{970} - H_{298})_{Mg_2SiO_4} = 104 \cdot 0 \text{ kJ}$$

$$(H_{970} - H_{298})_{Mg} = 28 \cdot 3 \text{ kJ}$$

$$(H_{970} - H_{298})_{Si} = 16 \cdot 4 \text{ kJ}$$

$$(H_{970} - H_{298})_{O_2} = 21 \cdot 7 \text{ kJ} \qquad \text{(Robie and Waldbaum 1968)}$$

Then:

$$\begin{aligned}
\Delta_f H_{Mg_2SiO_4}(1, 298) = {} & \Delta_f H_{Mg_2SiO_4}(1, 970) - (H_{970} - H_{298})_{Mg_2SiO_4} \\
& + 2(H_{970} - H_{298})_{Mg} + (H_{970} - H_{298})_{Si} \\
& + 2(H_{970} - H_{298})_{O_2} \\
= {} & -2186 \cdot 6 - 104 \cdot 0 + 2(28 \cdot 3) + 16 \cdot 4 + 2(21 \cdot 7) \\
= {} & -2174 \cdot 2 \text{ kJ}
\end{aligned}$$

Gibbs energy

The Gibbs energy, like the enthalpy, can only be defined with respect to a reference state, the one usually used being the elements in their most stable state at the pressure and temperature of interest. Thus, we refer to the Gibbs energy of formation from the elements, $\Delta_f G$. Then (A.1) should be written:

$$\Delta_f G(P, T) = \Delta_f H(1, T) - T\Delta_f S(1, T) + \int_1^P \Delta_f V(P, T) \, dP \qquad \text{(A.9)}$$

where the volume and entropy terms are also in terms of formation from the elements for consistency with the enthalpy and Gibbs energy. For example, for Mg_2SiO_4, forsterite:

$$\Delta_f S_{Mg_2SiO_4} = S_{Mg_2SiO_4} - 2S_{Mg} - S_{Si} - 2S_{O_2}$$

and

$$\Delta_f V_{Mg_2SiO_4} = V_{Mg_2SiO_4} - 2V_{Mg} - V_{Si} - 2V_{O_2}$$

Equation (A.9) can be rewritten using (A.7), (A.8) as:

$$\Delta_f G(P, T) = \Delta_f H(1, 298) - T\Delta_f S(1, 298) + \int_{298}^T \Delta_f C_p(1, T) \, dT$$

$$- T \int_{298}^T \frac{\Delta_f C_p(1, T)}{T} \, dT + \int_1^P \Delta_f V(P, T) \, dP \qquad \text{(A.10)}$$

Thus, the Gibbs energy of formation of a phase can be calculated from the 1 bar and 298 K values of the enthalpy of formation and the entropy, the heat capacity as a function of temperature and the volume as a function of both temperature and pressure plus these properties for the constituent elements of the phase. This rather unpleasant expression can be much simplified for the usual applications.

Simplifications

Usually we only require Gibbs energy values in order to calculate ΔG° values for balanced chemical reactions. The fact that the reaction is balanced with respect to each element results in the cancellation of many of the terms in (A.10). For example, consider the value of the change in the $(1, 298)$ entropy for the olivine–pyroxene–silicate liquid reaction:

$$SiO_2 + Mg_2SiO_4 = 2MgSiO_3$$
$$\text{liquid} \qquad \text{olivine} \qquad \text{orthopyroxene}$$

Assuming the $(1, 298)$ denotation on each of the terms, then:

$$\Delta S = 2\Delta_f S_{MgSiO_3} - \Delta_f S_{SiO_2} - \Delta_f S_{Mg_2SiO_4}$$

Expanding these entropies of formation:

$$\Delta S = 2(S_{MgSiO_3} - S_{Mg} - S_{Si} - \tfrac{3}{2}S_{O_2}) - (S_{SiO_2} - S_{Si} - S_{O_2})$$
$$- (S_{Mg_2SiO_4} - 2S_{Mg} - S_{Si} - 2S_{O_2})$$

The entropies of the elements now can be cancelled:

$$\Delta S = 2S_{MgSiO_3} - S_{SiO_2} - S_{Mg_2SiO_4}$$

This kind of cancellation can be applied to (A.10) as long as the result is only used in calculating ΔG° values and as long as the same cancellations are used in the Gibbs energy expression for the other end-members in a reaction. The resulting function, here called the Gibbs function, G', is not a Gibbs energy of formation, but has exactly the same effect when combined with other Gibbs function values in the calculation of ΔG°. Thus:

$$G'(P, T) = \Delta_f H(1, 298) - TS(1, 298) + \int_{298}^{T} C_p(1, T)\, dT$$

$$- T\int_{298}^{T} \frac{C_p(1, T)}{T}\, dT + \int_{1}^{P} V(P, T)\, dP \tag{A.11}$$

or:

$$\Delta G° = \Delta(\Delta_f H(1, 298)) - T\Delta S(1, 298) + \int_{298}^{T} \Delta C_p(1, T)\,dT$$

$$- T \int_{298}^{T} \frac{\Delta C_p(1, T)}{T}\,dT + \int_{1}^{P} \Delta V(P, T)\,dP$$

Thus $\Delta G°$ for a reaction can be calculated from values of:

(a) enthalpy of formation at $(1, 298)$;
(b) entropy at $(1, 298)$;
(c) heat capacity at 1 bar as a function of temperature. This is usually given in the form $C_p(1, T) = a + bT - c/T^2$ where a, b and c are constants which are different for each phase. The integrals in (A.11) are:

$$\int_{298}^{T} C_p(1, T)\,dT = a(T - 298) + b(T^2 - 298^2) + c\left(\frac{1}{T} - \frac{1}{298}\right)$$

(A.12)

$$\int_{298}^{T} \frac{C_p(1, T)}{T}\,dT = a \ln\left(\frac{T}{298}\right) + b(T - 298) + \frac{c}{2}\left(\frac{1}{T^2} - \frac{1}{298^2}\right)$$

(A.13)

(d) volume as a function of temperature and pressure.

Approximations

We can see what information is required to calculate $\Delta G°$ for a reaction from (A.11). However, we are in the unfortunate position of not often having all this information for end-members of minerals. In this section we will examine some of the approximations which can be made in calculating $\Delta G°$.

One way of evaluating the usefulness of an approximation is to consider the uncertainties introduced into our calculations by the likely magnitude of the error introduced by the approximation. Consider the equilibrium relation of a reaction which is going to be used as a geothermometer. As usual we have:

$$\Delta\mu = 0 = \Delta G° + RT \ln K$$

If, as is often the case, $\Delta G°$ is approximately linear in temperature and pressure then we may replace it by $A + BT + CP$, then:

$$0 = A + BT + CP + RT \ln K$$

If K is independent of temperature and pressure, then this can be re-arranged to give the equilibrium temperature:

$$T = \frac{-A - CP}{B + R \ln K}$$

For simplicity, we will consider the case when $K = 1$, then:

$$T = -\frac{A}{B} - \frac{C}{B} P \qquad\qquad (A.14)$$

In terms of a pressure–temperature diagram (figure A.3), the first term gives the temperature at zero pressure, the second determines the way the temperature changes with changing pressure. The uncertainty in the

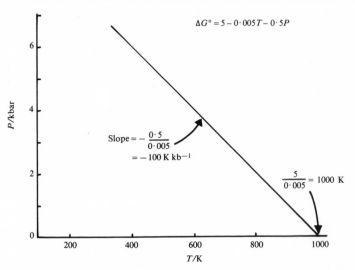

$\Delta G° = 5 - 0 \cdot 005 T - 0 \cdot 5 P$

$\text{Slope} = -\dfrac{0 \cdot 5}{0 \cdot 005}$

$= -100\ \text{K kb}^{-1}$

$\dfrac{5}{0 \cdot 005} = 1000\ \text{K}$

A.3 A P–T diagram with a reaction line illustrating the relationship between the position of the line and the equation $\Delta G° = A + BT + CP$ (i.e. the equilibrium relation with $K = 1$), for $A = 5$, $B = -0\cdot005$ and $C = -0\cdot5$.

position of the P–T line for a reaction depends on the uncertainties in the values of A, B and C for the reaction. For example, if $A = 5$ kJ and $B = -0\cdot005$ kJ K^{-1} giving a temperature at zero pressure of $5/0\cdot005 = 1000$ K, then an uncertainty of 1 kJ on A gives rise to a massive uncertainty of 200 K on the temperature (i.e. the temperature ranges from $(5+1)/0\cdot005 = 1200$ K to $(5-1)/0\cdot005 = 800$ K). On the other hand, if $A = 150$ kJ and $B = -0\cdot15$ kJ K^{-1} also giving 1000 K, then an uncertainty of 1 kJ on A only gives rise to an uncertainty of 7 K on the temperature. The major contribution to A is the sum of the enthalpies of formation of the end-members in the reaction. A method might be found for estimating the enthalpy of an end-member in a reaction which would be quite satisfactory in the calculation of A for the latter case while being hopelessly inadequate in the former case, if for example the probable error on the estimate was a few kilojoules. Thus when we start to consider approximations it is most important to realize that a method might be very useful for one category of reactions while being inadequate for another. The same is true for the acceptability of calorimetrically determined thermodynamic data. Excellent experimental precision on the enthalpy of formation of an end-member, for example $-2584 \cdot 5 \pm 0 \cdot 6$ kJ for sillimanite (Charlu *et al.* 1975), would not give a precise enough value of A for some reactions.

Reactions which only involve end-members of solid phases (solid–solid reactions) tend to have A and B values which are very small and frequently even good calorimetric data are too imprecise for the calculation of ΔG° for this type of reaction. On the other hand, reactions which involve solids and end-members of a fluid phase (solid–fluid reactions, for example, dehydration and decarbonation reactions) usually have large A and B values. Useful approximation reactions can be devised for the properties of end-members to be used in the calculation of ΔG° for this type of reaction. This subdivision between solid–solid and solid–fluid reactions is not clear-cut so that the equilibrium relation for each new reaction has to be considered on its merits.

The above logic concentrates on the sensitivity of the temperature at zero pressure to the value of A. Clearly this temperature is also sensitive to the value of B, the major contribution to B being the sum of the entropies of the end-members in the reaction. The position of the P–T line for the reaction at higher pressures is dependent on the uncertainties in B and C, the major contribution to C being the sum of the volumes of the end-members in the reaction. The approximations will be considered in turn, starting with the most straightforward, concerning the volume.

Volume

The volume expression which appears in ΔG° is:

$$\Delta \int_{1}^{P} V(P, T) \, dP = \int_{1}^{P} \Delta V(P, T) \, dP \tag{A.15}$$

For fluid end-members in the reaction the volume is a strong function of temperature and pressure, and so is the integral which appears in ΔG°. No satisfactory approximation has been devised for circumventing the necessity of measuring volumes over the range of temperatures and pressures of interest at least for CO_2 and H_2O. Values of the integral, in one form or another, are usually presented in tables, for example in Burnham *et al.* (1969) for H_2O. A new type of table is introduced later which simplifies the calculation of dehydration and decarbonation equilibria.

For end-members of solids in a reaction, the volume integral is a weak function of temperature and pressure. For constant values of α and κ (A.5):

$$\int_1^P V(P, T)\, dP = PV(1, 298)[1 + \alpha(T - 298) - \kappa P/2]$$

However the effect of α and κ for the end-members will tend to cancel on (A.15) because the values of α and κ for each of the end-members (of solids) are of the same size and magnitude. Nevertheless, following the logic developed on p. 253, if the value of B is very small, then even the small difference in C contributed by α and κ will have a significant effect on the position of the P–T line for the reaction. The main situation in which it is necessary to consider α and κ is for solid–solid reactions, particularly polymorphic transitions like diamond–graphite.

Normally, particularly for solid–fluid reactions, the effect of α and κ on the volume integral term is quite negligible. This means that $V(P, T)$ can be approximated by $V(1, 298)$ (equation (A.6)):

$$\int_1^P V(P, T)\, dP = PV(1, 298)$$

For solid–solid reactions for which this approximation is reasonable:

$$\Delta \int_1^P V(P, T)\, dP = P \Delta V(1, 298)$$

For solid–fluid reactions, the volume integral for fluid end-members can be separated from the volume integrals for the solid end-members. For example, for a reaction involving n H_2O molecules and m CO_2 molecules:

$$\Delta \int_1^P V(P, T)\, dP = P \Delta V(1, 298)_s + n \int_1^P V_{H_2O}(P, T)\, dP + m \int_1^P V_{CO_2}(P, T)\, dP$$

$$(A.16)$$

where the first term on the right-hand side refers to the sum of the volume integrals for the solid end-members, hence the subscript S.

The availability of volume data for solids is not usually a problem because the data can be generated simply from the X-ray diffraction measurement of unit cell dimensions. The main data tabulation is Robie *et al.* (1967). On the other hand, α and κ values are available for relatively few solids of geological interest, for example Clark (1966).

Heat capacity
The heat capacity expression which appears in ΔG° is:

$$\int_{298}^{T} \Delta C_p(1, T)\, dT - T \int_{298}^{T} \frac{\Delta C_p(1, T)}{T}\, dT \qquad (A.17)$$

High temperature heat capacities have been measured for relatively few end-members of geological interest. Further, the published heat capacities were not always obtained on well characterized samples and so some are of doubtful validity. There are two approaches to approximation here:

(1) devise a method for estimating heat capacities;
(2) devise a method for calculating ΔG° in which the heat capacities are not required.

It was recognized long ago that heat capacities are approximately additive—by which it is meant that, for example the heat capacity of Mg_2SiO_4 can be approximated by adding the heat capacities of $2(MgO)$ and SiO_2. The amount by which this approximation is in error appears to be related to the difference between the volume of the end-member and the sum of the volumes of the component oxides, and various volume corrections have been suggested. Consider Mg_2SiO_4, forsterite. The measured heat capacity can be expressed as

$$C_{p,Mg_2SiO_4} = 0 \cdot 1498 + 2 \cdot 74 \times 10^{-5} T - 3565/T^2 \text{ kJ} \qquad (A.18)$$

The heat capacities of the oxides are:

$$C_{p,MgO} = 0 \cdot 0426 + 0 \cdot 73 \times 10^{-5} T - 619/T^2 \text{ kJ}$$
$$C_{p,SiO_2} = 0 \cdot 0469 + 3 \cdot 43 \times 10^{-5} T - 1130/T^2 \text{ kJ}$$

So the oxide sum heat capacity is:

$$C_{p,Mg_2SiO_4}^{sum} = 0 \cdot 1321 + 4 \cdot 89 \times 10^{-5} T - 2368/T^2 \qquad (A.19)$$

Comparing the value of the heat capacity integral (A.17) for Mg_2SiO_4 at 1000 K using (A.18) and (A.19), we have $-73 \cdot 1$ kJ and $-72 \cdot 7$ kJ—very good agreement for most purposes. This is partly because $V_{Mg_2SiO_4} = 4 \cdot 379$ kJ kbar^{-1} is very similar to the oxide sum volume $V_{Mg_2SiO_4}^{sum} = 4 \cdot 519$ kJ kbar^{-1}. If this difference is much larger then the discrepancy between estimated and measured C_p values becomes larger and a volume correction is required.

The approximate additivity of heat capacities means that unknown heat capacities can be estimated. However, this additivity leads to a method of calculating $\Delta G°$ in which heat capacities are not required. If this additivity applied to each end-member in a balanced reaction, then an obvious approximation is $\Delta C_p = 0$ for the reaction, because the reaction is balanced for the oxides. This would mean that the heat capacity integrals (A.17) are zero. However, the additivity only works on end-members of solid phases, and so $\Delta C_p = 0$ is only likely to work on solid–solid reactions. Usually this is of little help because, as already stated above, approximations are usually inadequate for solid–solid reactions. If A and B are indeed large for a solid–solid reaction then, using this approximation and the volume approximation, we have:

$$\Delta G° = \Delta(\Delta_f H(1, 298)) - T\Delta S(1, 298) + P\Delta V(1, 298) \qquad (A.20)$$

which has the merit of great simplicity. It can also be compared directly with the expression of $\Delta G°$ as $A + BT + CP$ used earlier, with in this case:

$$A = \Delta(\Delta_f H(1, 298)), \qquad B = -\Delta S(1, 298) \quad \text{and} \quad C = \Delta V(1, 298)$$

If we assume that the $\Delta C_p = 0$ approximation contributes an uncertainty of 2 kJ to A and 0.005 kJ K^{-1} to B, we can see if the A and B values calculated using $\Delta C_p = 0$ for a particular solid–solid reaction are likely to be useful.

Consider the reaction:

$$3CaAl_2Si_2O_8 = 2Al_2SiO_5 + Ca_3Al_2Si_3O_{12} + SiO_2$$

$$\text{plagioclase} \qquad \text{kyanite} \qquad \text{garnet} \qquad \text{quartz}$$

Using the data at the end of this appendix and (A.20):

$$\begin{aligned}
A \;\; &= \Delta(\Delta_f H(1, 298)) = 2(-2592) + (-6646) + (-910.6) - 3(-4230) \\
&= -50.6 \text{ kJ} \\
-B &= \Delta S(1, 298) = 2(0.0853) + (0.2414) + (0.0413) - 3(-4230) \\
&= -0.1548 \text{ kJ K}^{-1} \\
C \;\; &= \Delta V(1, 298) = 2(4.409) + (12.530) + (2.269) - 3(10.079) \\
&= -6.62 \text{ kJ kbar}^{-1}
\end{aligned}$$

Assuming $\Delta C_p = 0$ is a reasonable approximation, then a condition for equilibrium of plagioclase–kyanite–garnet–quartz is given by:

$$\begin{aligned}
\Delta\mu = 0 &= \Delta G° + RT \ln K \\
&= -50.6 + 0.1548T - 6.62P + RT \ln K
\end{aligned}$$

For $K = 1$ this defines a P–T line, $T = -50.6/0.1548 - (6.62/0.1548)P = 327 + 42.8P$ with T in K and P in kbar. The suggested uncertainties in A and B introduced by using $\Delta C_p = 0$ contribute an uncertainty of less than 25 K to the calculated temperature—a not unreasonable uncertainty for most purposes.

Consider, however, the reaction:

$$2MgSiO_3 = Mg_2SiO_4 + SiO_2$$

orthopyroxene   olivine   quartz

Using the data in the appendix, and assuming $\Delta C_p = 0$:

$$A = \Delta(\Delta_f H(1, 298)) = 6 \cdot 0 \text{ kJ}$$
$$-B = \Delta S(1, 298) = -0 \cdot 0007 \text{ kJ K}^{-1}$$
$$C = \Delta V(1, 298) = 0 \cdot 36 \text{ kJ kbar}^{-1}$$

In this case, A and particularly B are too small for the $\Delta C_p = 0$ approximation to be useful. Further the entropy (and the enthalpy) data in the appendix are unlikely to be precise enough to define the position of the reaction, even approximately, even including heat capacities in the calculation. The only reasonable way to perform calculations on this type of reaction is to use an experimental determination of the reaction directly, as shown on p. 148.

The reason why the $\Delta C_p = 0$ approximation does not work for dehydration reactions is because the heat capacity contribution to ΔC_p is different for H_2O fluid and structurally bound H_2O in a solid. The same applies to the fluid end-member in other solid–fluid reactions. However the heat capacity difference between the free and structurally bound states appears to be more or less the same for different reactions, being about $-0 \cdot 013$ kJ per H_2O molecule in the reaction and $-0 \cdot 009$ kJ per CO_2 in the reaction. If we call this heat capacity difference r, then the heat capacity integral (A.17) becomes:

$$\int_{298}^{T} r\, dT - T \int_{298}^{T} \frac{r}{T}\, dT = r(T - 298) - rT \ln\left(\frac{T}{298}\right)$$

For a reaction involving n H_2O molecules and m CO_2 molecules, using this approximation and the volume approximation (A.16), we have:

$$\Delta G^\circ = \Delta(\Delta_f H(1, 298)) - T\Delta S(1, 298) + P\Delta V(1, 298)_s$$

$$+ n\left(r_{H_2O}\left(T - 298 - T \ln\left(\frac{T}{298}\right)\right) + \int_{1}^{P} V_{H_2O}(P, T)\, dP\right)$$

$$+ m\left(r_{CO_2}\left(T - 298 - T \ln\left(\frac{T}{298}\right)\right) + \int_{1}^{P} V_{CO_2}(P, T)\, dP\right) \qquad (A.21)$$

The terms in large parentheses are functions of temperature and pressure and need to be tabulated as such. We can write:

$$F_i = r_i\left(T - 298 - T \ln\left(\frac{T}{298}\right)\right) + \int_1^P V_i(P, T)\, dP$$

and then rewrite (A.21) as:

$$\Delta G^\circ = \Delta(\Delta_f H(1, 298)) - T\Delta S(1, 298) + P\Delta V(1, 298)_s + nF_{H_2O} + mF_{CO_2} \tag{A.22}$$

F_{H_2O} and F_{CO_2} can be tabulated for a range of temperatures and pressures. Both F_{H_2O} and F_{CO_2} are more or less linear over a temperature range of 100 or 200 K at a particular pressure and this allows a neat way of calculating ΔG° for reactions involving H_2O and/or CO_2. Suppose we know that a reaction takes place in a particular temperature range for a particular pressure. Now F can be expressed as a linear function of temperature for this temperature range at this pressure, thus:

$$F_{H_2O} = a'_{H_2O} + b'_{H_2O}T \quad \text{and} \quad F_{CO_2} = a'_{CO_2} + b'_{CO_2}T \tag{A.23}$$

Combining (A.22) and (A.23):

$$\Delta G^\circ = [\Delta(\Delta_f H(1, 298)) + P\Delta V(1, 298) + na'_{H_2O} + ma'_{CO_2}]$$
$$+ T[-\Delta S(1, 298) + nb'_{H_2O} + mb'_{CO_2}] \tag{A.24}$$

For example, consider the reaction

$$KAl_3Si_3O_{10}(OH)_2 + SiO_2 = KAlSi_3O_8 + Al_2SiO_5 + H_2O$$

muscovite   quartz   feldspar   andalusite   fluid

Using the data in the appendix:

$$\Delta(\Delta_f H(1, 298)) = 98\cdot0 \text{ kJ}$$
$$\Delta S(1, 298) = 0\cdot1747 \text{ kJ K}^{-1}$$
$$\Delta V(1, 298) = -0\cdot31 \text{ kJ kbar}^{-1}$$

For application of (A.24), $n = 1$ and $m = 0$ for this reaction. If we want the equilibrium temperature at 2 kbar and with $K = 1$ for the reaction, then:

$$\Delta G^\circ = 98\cdot0 - 2(0\cdot31) + a'_{H_2O} + T(-0\cdot1747 + b'_{H_2O}) = 0$$

The equilibrium temperature is somewhere near 600 °C. For a pressure of 2 kbar and for the temperature range 873 to 973 K, $a'_{H_2O} = -25\cdot7$ kJ and $b'_{H_2O} = 0\cdot0927$ kJ K^{-1}. Therefore:

$$\Delta G^\circ = 98\cdot0 - 2(0\cdot31) - 25\cdot7 + T(-0\cdot1747 + 0\cdot0927) = 0$$

or $71\cdot7 - 0\cdot0820T = 0$ which gives $T = 874$ K $= 601$ °C.

For solid–fluid reactions other than decarbonation and dehydration reactions the situation is more complicated. For example, there is no single value of r for O_2 in oxidation reactions. However most of the calculations on solid–fluid reactions will involve H_2O and/or CO_2. For those involving other fluid end-members, for example O_2 and S_2, the ΔG° of the reaction will be given in the form $A + BT + CP$ and the activity of the fluid end-member will include the fugacity of the end-member.

Entropy and enthalpy

In the last section, (A.20) and (A.22) summarize the results of approximations which are suitable for calculating ΔG° for some solid–solid reactions and most solid–fluid reactions involving H_2O and/or CO_2. The main unknowns in these expressions are $\Delta_f H(1, 298)$ and $S(1, 298)$ values for each end-member in the reaction. Approximations and methods of calculation for these will be considered first. An approach for dealing with reactions with small values of A and B like the orthopyroxene–olivine–quartz reaction above, will be considered later.

Recalling that the entropy is defined in terms of a heat capacity integral (A.7), and noting the additivity of the heat capacities, we might expect entropies to be additive as well. In fact this is the basis of the main method of estimating entropies. One problem is that the additivity of heat capacities starts to break down at very low temperatures, and as the entropy is a result of integrating C_p/T from 0 K to the temperature of interest, this means that the additivity approach is less successful for entropies than heat capacities at high temperatures. However a volume correction has been devised which substantially improves the oxide sum entropy. The method involves:

$$S_i = S_i^{\text{sum}} + 0 \cdot 02(V_i - V_i^{\text{sum}}) \tag{A.25}$$

For example for Mg_2SiO_4, forsterite, the measured entropy at $(1, 298)$ from Robie and Waldbaum (1968) is $S_{Mg_2SiO_4}(1, 298) = 0 \cdot 0952 \text{ kJ K}^{-1}$, while $S_{Mg_2SiO_4}^{\text{sum}} = 2(0 \cdot 0269) + (0 \cdot 0413) = 0 \cdot 0951 \text{ kJ K}^{-1}$. The volume correction is $0 \cdot 02[4 \cdot 379 - (2(1 \cdot 125) + (2 \cdot 269))] = -0 \cdot 0028 \text{ kJ K}^{-1}$, giving an estimated value of the entropy of $0 \cdot 0923 \text{ kJ K}^{-1}$, which is 2% lower than the calorimetric value.

The entropy and volume values for H_2O, ice, are used to estimate entropies of hydrous phases. The values of the entropy and volume for some oxides which can be used for estimating entropies using (A.25) are given in table A.1.

No satisfactory equivalent approach has been devised for calculating enthalpies of formation, except for, perhaps, clay minerals (for example, Tardy and Garrels 1974). The main method for estimating enthalpies of formation uses the experimentally determined position of pressure–temperature lines for reactions. As the position of a P–T line is very sensitive to the enthalpies of formation of the end-members in a reaction

Table A.1 Values of $S(1, 298)$ and $V(1, 298)$ for the oxides which can be used to estimate $(1, 298)$ entropy values using (A.25). The values are from Robie and Waldbaum (1968). The value of V_{Na_2O} was calculated using (A.25) re-arranged and the known entropy of nepheline.

Oxide	$S(1, 298)/$ $kJ\,K^{-1}$	$V(1, 298)/$ $kJ\,kbar^{-1}$
SiO_2	0·0413	2·269
Al_2O_3	0·0510	2·558
TiO_2	0·0504	1·882
Fe_2O_3	0·0874	3·027
FeO	0·0608	1·200
MgO	0·0269	1·125
MnO	0·0597	1·322
CaO	0·0397	1·676
Na_2O	0·0753	1·750
K_2O	0·0941	4·038
H_2O	0·0447	1·964

the position of the line can be utilized to obtain precise estimates of enthalpies of reaction, if all the appropriate entropies and volumes are known. This can be seen most clearly with the help of an example.

Consider again the plagioclase–kyanite–garnet–quartz equilibrium on p. 129. The reaction is:

$$3CaAl_2Si_2O_8 = 2Al_2SiO_5 + Ca_3Al_2Si_3O_{12} + SiO_2$$

    plagioclase   kyanite   garnet   quartz

For which:

$$\Delta G° = -50·6 + 0·1548T - 6·62P$$

with T in K and P in kbar. If $K = 1$, the equilibrium condition for this reaction is:

$$0 = -50·6 + 0·1548T - 6·62P \tag{A.26}$$

Imagine that the A value was imperfectly known, but that the reaction was known to lie between 750 K and 760 K at 10 kbar from high temperature–pressure experimental work. Rewriting (A.26) as $0 = A + 0·1548T - 6·62P$, we can calculate A so that it is consistent with the experiments. The two extreme values of A consistent with the extremes of the experimental bracket are $-0·1548(750) + 6·62(10) = -49·9$ kJ and $-0·1548(760) + 6·62(10) = -51·3$ kJ, a range of only 1·4 kJ. If we assume that we know the enthalpies of formation of kyanite, anorthite and quartz we can calculate the enthalpy of formation of grossular:

$$\Delta_f H_{Ca_3Al_2Si_3O_{12}} = A - 2\Delta_f H_{Al_2SiO_5} - \Delta_f H_{SiO_2} + 3\Delta_f H_{CaAl_2Si_2O_8}$$

The enthalpy of formation of grossular lies between $-6645\cdot3$ kJ and $-6646\cdot7$ kJ. If the experimental bracketing of the position of a reaction has been performed at a number of pressures, then a $\Delta_f H$ bracket can be calculated for each of these pressures. The favoured enthalpy of formation is then the one which lies centrally within the overlap of all the $\Delta_f H$ brackets.

The calculation of enthalpies of formation from the experimentally determined P–T lines for reactions assumes that entropies and volumes are known, as well as a set of enthalpy of formation values. With these, it is possible to progress through the available experimental studies building up a set of thermodynamic data which is consistent with all the experiments (as long as all the experimental studies are consistent with each other!). This approach is hazardous, due to uncertainties in entropy values, uncertainties in the reliability of experimental studies (particularly when a reaction has been studied several times with inconsistent results), and so on. However, the next section in this appendix contains a data set which is an attempt at providing an internally consistent set. Some notes after the tables provide information on the reactions used in generating data, and some of the problems encountered.

An example will illustrate the approach taken. Consider the calculation of the enthalpy of formation of muscovite. Chatterjee and Johannes (1974) published a detailed study of mineral equilibria involving muscovite, including bracketing the reactions:

(a) $\underset{\text{muscovite}}{KAl_3Si_3O_{10}(OH)_2} = \underset{\text{feldspar}}{KAlSi_3O_8} + \underset{\text{corundum}}{Al_2O_3} + \underset{\text{fluid}}{H_2O}$

(b) $\underset{\text{muscovite}}{KAl_3Si_3O_{10}(OH)_2} + \underset{\text{quartz}}{SiO_2} = \underset{\text{feldspar}}{KAlSi_3O_8} + \underset{\text{andalusite}}{Al_2SiO_5} + \underset{\text{fluid}}{H_2O}$

They present temperature brackets for both reactions over a range of temperatures. If we assume that we know all the entropies and volumes, and the enthalpies of formation of everything but muscovite, then using the data in the appendix (except using the previously accepted enthalpy of formation of andalusite of $-2591\cdot5$ kJ):

$$\Delta G° = -5873\cdot9 - \Delta_f H_{KAl_3Si_3O_{10}(OH)_2} - 0\cdot1738T - 0\cdot632P$$
$$\Delta G° = -5879\cdot5 - \Delta_f H_{KAl_3Si_3O_{10}(OH)_2} - 0\cdot1747T - 0\cdot312P$$

for reactions (a) and (b) respectively, with T in K and P in kbar.

To see the method of calculation, we take the 2 kbar bracket, 913 K–933 K for reaction (a). For this temperature range:

$$F_{H_2O} = -25\cdot7 + 0\cdot0927T$$

Using (A.24) and noting that the experimentally determined reaction is the same as the end-member reaction, so that $K = 1$, then:

$$\Delta G° = -5873\cdot9 - \Delta_f H_{KAl_3Si_3O_{10}(OH)_2} - 2(0\cdot632) - 25\cdot7$$
$$+ T(-0\cdot1738 + 0\cdot0927) = 0$$

Re-arranging:

$$\Delta_f H_{KAl_3Si_3O_{10}(OH)_2} = -5900\cdot7 - 0\cdot0811T \text{ kJ}$$

So the enthalpy of formation of muscovite lies between $-5900\cdot7 - 0\cdot0811(913) = -5974\cdot7$ kJ and $-5900\cdot7 - 0\cdot0811(933) = -5976\cdot4$ kJ. This procedure has been repeated for each of the experimental brackets and the results are presented in table A.2.

Table A.2 Values of the enthalpy of formation of muscovite at $(1, 298)$ which make the appropriate reaction pass through each end of the experimental brackets of Chatterjee and Johannes (1974) for (a) and (b).

Reaction (a)			Reaction (b)		
Pressure/ kbar	Temperature/K of exper- imental brackets	Enthalpy/ kJ	Pressure/ kbar	Temperature/K of exper- imental brackets	Enthalpy/ kJ
1	873	−5974·5	0·5	793	−5976·9
	903	−5977·2		833	−5980·0
2	913	−5974·7	1	823	−5976·2
	933	−5976·4		843	−5978·0
4	963	−5974·2	2	863	−5976·5
	983	−5975·7		878	−5977·7
6	1013	−5974·2	3	893	−5976·3
	1023	−5975·0		913	−5977·0
8	1053	−5973·8	4	933	−5977·1
	1073	−5975·3		944	−5977·9
			5	963	−5977·3
				978	−5978·4

These enthalpy of formation brackets are plotted against pressure in figure A.4. The enthalpy of formation should not depend on pressure as it is a $(1, 298)$ value. A dependence of the enthalpy of formation brackets on pressure would indicate that the $\Delta S(1, 298)$ or the $\Delta V(1, 298)S$, for the reaction was incorrect. There is no evidence for this for either of the reactions. The preferred value of the enthalpy of reaction is one which lies within the overlap of all the enthalpy brackets. For reaction (a) it is -5975 kJ, while for (b) it is $-5977\cdot5$ kJ. What is the cause of this discrepancy? Some data were assumed to be correct in calculating the brackets in figure A.4. However muscovite and sanidine appear in the same proportion in both reactions, so that changing any of the properties of these will change the position of all the brackets by more or less the same amount so not affecting the discrepancy. The obvious culprit is the enthalpy of formation of andalusite, as the enthalpies of formation of quartz and corundum are reasonably well known. Further, the tabulated value of $\Delta_f H_{Al_2SiO_5}(1, 298)$ for andalusite is $-2591\cdot5 \pm 3\cdot0$ kJ in Robie and Waldbaum (1968), involving a substantial uncertainty. The enthalpy of formation of andalusite has to be adjusted by only $2\cdot5$ kJ to make the enthalpy of formation of muscovite calculated from reaction (b) consistent with the value calculated from reaction (a), -5975 kJ. This adjustment of the andalusite value is within the stated uncertainty of the original calorimetric work.

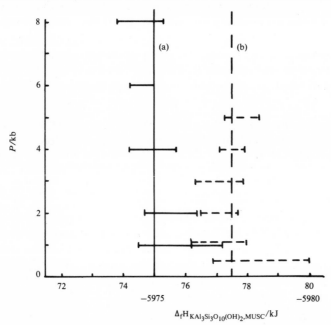

A.4 A P–$\Delta_f H_{KAl_3Si_3O_{10}(OH)_2,muscovite}$ diagram showing the enthalpy brackets which correspond to the temperature brackets determined experimentally at each pressure, for reactions (a) and (b) on p. 249 using the values given in table A.2.

Having calculated this value for muscovite and adjusted the value for andalusite, it is now possible to consider other reactions which involve muscovite and some other end-member for which the enthalpy of formation is not available, and so on. It is helpful to have some experimentally determined reactions which have not been used to estimate enthalpy values to check the internal consistency of the data set. It is important to realize that many of the estimated enthalpy values in the data set are dependent directly or indirectly on many of the other values. This means that updating the data set with new calorimetric or experimental data usually involves starting at the beginning again.

The above calculations use the approximations on the pressure-temperature dependence of the volume and on the heat capacities. As already stated, these approximations are not adequate for calculating $\Delta G°$ for most solid–solid reactions. In certain cases it is possible to a follow a rigorous approach in the calculation of a P–T line for a reaction, but this is possible only when heat capacities, thermal expansions, isothermal compressibilities, entropies, enthalpies of formation and volumes are well known! However, in most cases, a more direct approach is the only feasible way. This approach uses the experimentally determined position of a reaction to calculate $\Delta G°$ for that reaction. This is only going to be of any use if the equilibrium is required for conditions other than those covered by the experiments; in other words, if the results of the experiments need to be extrapolated, particularly in composition, for the application of interest. As the P–T lines of such solid–solid reactions are usually more or less straight, it is usually assumed that $\Delta G°$ is linear in temperature and pressure and so can be written as usual as $\Delta G° = A + BT + CP$.

Consider the orthopyroxene (opx)–olivine (ol)–quartz (q) assemblage found in rhyolites and granites. One end-member reaction corresponding to this assemblage is:

$$2FeSiO_3 = Fe_2SiO_4 + SiO_2$$
$$\quad\text{opx}\qquad\quad\text{ol}\qquad\text{q}$$

There are insufficient thermodynamic data to calculate $\Delta G°$ for this reaction. Smith (1971) has however determined the approximate position of the end-member reaction; it goes near 1473 K at 17·5 kbar and 773 K at 9·5 kbar. This can be expressed in the form $P = u + vT$, by using the method for putting a line through two points:

$$\frac{T - 773}{1573 - 773} = \frac{P - 9·5}{17·5 - 9·5}$$

Re-arranging gives:

$$P = 0·01143T + 0·6657 \qquad \text{with } P \text{ in kbar and } T \text{ in K} \tag{A.27}$$

Now, the equilibrium relation applicable to these experiments is:

$$\Delta G° = A + BT + CP = 0 \quad \text{or} \quad P = -\frac{B}{C}T - \frac{A}{C}$$

Comparing with the equation for the experimentally determined P–T line:

$$A = -0·6657C \text{ kJ}$$
$$B = -0·01143C \text{ kJ K}^{-1} \tag{A.28}$$

Now, as we are interested in relatively low pressures, we can ignore α and κ, so that from (A.20), $C = \Delta V(1, 298) = 0·308 \text{ kJ kbar}^{-1}$ using data in

Robie *et al.* (1967). Thus (A.28) becomes:

$A = -0.205$ kJ

$B = -0.00352$ kJ K^{-1}

and:

$$\Delta G^\circ = -0.205 - 0.00352T + 0.308P$$

The equilibrium relation for the reaction is

$$\Delta \mu = 0 = \Delta G^\circ + RT \ln K$$

$$= -0.205 - 0.00352T + 0.308P + RT \ln \frac{a_{Fe_2SiO_4,ol}}{a^2_{FeSiO_3,opx}} \quad \text{(A.29)}$$

Activities of the end-members in the olivine and orthopyroxene of a mineral assemblage can now be inserted into this equation to discover the P-T line on which the assemblage could have formed. It is worth noting that the position of the calculated P-T line is very sensitive to the formulation of the activity–composition relations of the olivine and orthopyroxene, because A and B are so small. For example, consider an olivine–orthopyroxene–quartz assemblage, with $a_{Fe_2SiO_4} = 0.7$ and $a_{FeSiO_3} = 0.67$ so that $K = 0.7/0.67^2$. Applying this in (A.29) gives a temperature of 1179 K. However, including a trivial 1% uncertainty on each of the activities contributes a massive uncertainty of over 1000 K to the calculated temperature.

Thermodynamic data

For most dehydration and decarbonation reactions and a few solid–solid reactions (A.22) is applicable:

$$\Delta G^\circ = \Delta(\Delta_f H(1, 298) - T\Delta S(1, 298) + P\Delta V(1, 298)_s + nF_{H_2O} + mF_{CO_2}$$
$$\text{(A.30)}$$

for a reaction involving n H_2O and m CO_2 molecules. For solid–solid reactions $n = m = 0$, for dehydration reactions $m = 0$, and for decarbonation reactions $n = 0$. The thermodynamic data required are values of $\Delta_f H(1, 298)$, $S(1, 298)$ and $V(1, 298)$ for end-members of phases (table A.3), plus tables of F_{H_2O} and F_{CO_2}. The F tables are given as coefficients a' and b' where $F = a' + b'T$ (table A.4). An example of the calculation of ΔG° using these tables is given on p. 129.

The tables of data are an attempt at providing a set of thermodynamic data which is consistent with available experimental work on the position of P-T lines for reactions and calorimetric work (up to May 1st, 1977). Most of the values in the data set are dependent on other values via the position of P-T lines for reactions. This means that updating the data set with new calorimetric or experimental data usually involves adjusting many values in the data set.

Table A.3 Thermodynamic data for end-members of phases. All volumes from Robie *et al.* (1967) and entropies and enthalpies from Robie and Waldbaum (1968) unless otherwise indicated in the notes to this table.

Name	Formula	$\Delta_f H(1, 298)/$ kJ	$S(1, 298)/$ kJ K^{-1}	$V(1, 298)/$ kJ kbar^{-1}	Note
H$_2$O(gas)	H$_2$O	$-241 \cdot 8$	$0 \cdot 1887$	—	
CO$_2$(gas)	CO$_2$	$-393 \cdot 5$	$0 \cdot 2136$	—	
Corundum	Al$_2$O$_3$	$-1675 \cdot 3$	$0 \cdot 0510$	$2 \cdot 558$	
Periclase	MgO	$-601 \cdot 7$	$0 \cdot 0269$	$1 \cdot 125$	
Quartz	SiO$_2$	$-910 \cdot 6$	$0 \cdot 0413$	$2 \cdot 269$	
Rutile	TiO$_2$	$-944 \cdot 6$	$0 \cdot 0504$	$1 \cdot 882$	
Diaspore	AlO(OH)	$-998 \cdot 3$	$0 \cdot 0353$	$1 \cdot 776$	1
Gibbsite	Al(OH)$_3$	$-1294 \cdot 2$	$0 \cdot 0701$	$3 \cdot 196$	2
Brucite	Mg(OH)$_2$	$-927 \cdot 0$	$0 \cdot 0631$	$2 \cdot 463$	3
Calcite	CaCO$_3$	$-1207 \cdot 5$	$0 \cdot 0927$	$3 \cdot 693$	
Dolomite	CaMg(CO$_3$)$_2$	$-2327 \cdot 0$	$0 \cdot 1552$	$6 \cdot 434$	17
Magnesite	MgCO$_3$	$-1112 \cdot 1$	$0 \cdot 0657$	$2 \cdot 802$	4
Andalusite	Al$_2$SiO$_5$	$-2589 \cdot 0$	$0 \cdot 0932$	$5 \cdot 147$	9
Kyanite	Al$_2$SiO$_5$	$-2592 \cdot 0$	$0 \cdot 0853$	$4 \cdot 409$	9
Sillimanite	Al$_2$SiO$_5$	$-2582 \cdot 7$	$0 \cdot 0993$	$4 \cdot 984$	9
Sphene	CaTiSiO$_5$	$-2595 \cdot 7$	$0 \cdot 1292$	$5 \cdot 565$	5
Spinel	MgAl$_2$O$_4$	$-2299 \cdot 1$	$0 \cdot 0806$	$3 \cdot 971$	7
Albite (high)	NaAlSi$_3$O$_8$	$-3915 \cdot 5$	$0 \cdot 2287$	$10 \cdot 043$	6
Anorthite	CaAl$_2$Si$_2$O$_8$	$-4230 \cdot 0$	$0 \cdot 2027$	$10 \cdot 079$	14
Sanidine	KAlSi$_3$O$_8$	$-3956 \cdot 8$	$0 \cdot 2228$	$10 \cdot 893$	6
Forsterite	Mg$_2$SiO$_4$	$-2174 \cdot 2$	$0 \cdot 0952$	$4 \cdot 379$	7
Enstatite (ortho)	MgSiO$_3$	$-1545 \cdot 4$	$0 \cdot 0679$	$3 \cdot 144$	12
Diopside	CaMgSi$_2$O$_6$	$-3205 \cdot 0$	$0 \cdot 1431$	$6 \cdot 609$	18
Wollastonite	CaSiO$_3$	$-1630 \cdot 0$	$0 \cdot 0820$	$3 \cdot 993$	8
Clinochrysotile	Mg$_3$Si$_2$O$_5$(OH)$_4$	$-4361 \cdot 5$	$0 \cdot 2213$	$10 \cdot 715$	13
Kaolinite	Al$_2$Si$_2$O$_5$(OH)$_4$	$-4121 \cdot 0$	$0 \cdot 2030$	$9 \cdot 952$	15
Margarite	CaAl$_4$Si$_2$O$_{10}$(OH)$_2$	$-6240 \cdot 8$	$0 \cdot 2636$	$12 \cdot 964$	16
Muscovite	KAl$_3$Si$_3$O$_{10}$(OH)$_2$	$-5975 \cdot 0$	$0 \cdot 2887$	$14 \cdot 081$	9
Paragonite	NaAl$_3$Si$_3$O$_{10}$(OH)$_2$	$-5946 \cdot 0$	$0 \cdot 2721$	$13 \cdot 210$	10
Phlogopite	KMg$_3$AlSi$_3$O$_{10}$(OH)$_2$	$-6217 \cdot 0$	$0 \cdot 3197$	$14 \cdot 991$	22
Pyrophyllite	Al$_2$Si$_4$O$_{10}$(OH)$_2$	$-5639 \cdot 5$	$0 \cdot 2368$	$12 \cdot 590$	11
Clinochlore	Mg$_5$Al$_2$Si$_3$O$_{10}$(OH)$_8$	$-8886 \cdot 0$	$0 \cdot 4561$	$20 \cdot 754$	
Grossular	Ca$_3$Al$_2$Si$_3$O$_{12}$	$-6646 \cdot 0$	$0 \cdot 2414$	$12 \cdot 530$	20
Talc	Mg$_3$Si$_4$O$_{10}$(OH)$_2$	$-5899 \cdot 0$	$0 \cdot 2608$	$13 \cdot 625$	12
Tremolite	Ca$_2$Mg$_5$Si$_8$O$_{22}$(OH)$_2$	$-12320 \cdot 0$	$0 \cdot 5489$	$27 \cdot 292$	19
Clinozoisite	Ca$_2$Al$_3$Si$_3$O$_{12}$(OH)	$-6904 \cdot 5$	$0 \cdot 2841$	$13 \cdot 648$	21

Notes to Table A.3

1 Enthalpy of formation of diaspore calculated from experiments on the diaspore–corundum–H$_2$O reaction by Haas (1972).
2 Enthalpy of formation of gibbsite from Gross *et al.* and Hemingway and Robie, see Thompson (1974).
3 Enthalpy of formation of brucite calculated from experiments on the brucite–periclase–

H_2O reaction by Barnes and Ernst (1963). The value, $-927\cdot0$ kJ, is within the uncertainty of the value given in Robie and Waldbaum (1968), $-925\pm2\cdot9$ kJ.

4 Enthalpy of formation of magnesite calculated from experiments on the magnesite–periclase–CO_2 reaction by Harker and Tuttle (1955). The value, $-1112\cdot1$ kJ is within the uncertainty of the value given in Robie and Waldbaum (1968), $-1113\cdot3\pm1\cdot3$ kJ.

5 Enthalpy of formation of sphene calculated from experiments on the sphene–calcite–rutile–quartz–CO_2 reaction by Hunt and Kerrick (1977). The value is $4\cdot5$ kJ larger than the value in Robie and Waldbaum (1968).

6 The thermodynamic data for sanidine are from Chatterjee and Johannes (1974), and the enthalpy of formation of albite is corrected in the same way as the sanidine value for the systematic errors on the value in Robie and Waldbaum (1968).

7 The enthalpies of formation of forsterite and spinel are taken from Charlu et al. (1975).

8 Enthalpy of formation of wollastonite calculated from experiments on the wollastonite–calcite–quartz–CO_2 reaction by Harker and Tuttle (1956) and Greenwood (1967). The value, -1630 kJ, is just outside the uncertainty on the value given in Robie and Waldbaum (1968), $-1634\cdot4\pm3\cdot6$ kJ.

9 The calculation of the enthalpy of formation of muscovite was considered on p. 249. As stated there, this necessitates a correction of $2\cdot5$ kJ to the enthalpy of formation of andalusite. The thermodynamic data for sillimanite and kyanite were then adjusted to give agreement with the Richardson et al. (1969) aluminosilicate phase diagram. This was done using (A.30) so that in diagrams like figure 3.22 a reaction involving one aluminosilicate would intersect the same reaction but with another of the aluminosilicates at a point on the appropriate P–T line of the aluminosilicate phase diagram. Unfortunately the use of (A.30) results in the P–T line for the andalusite–sillimanite reaction being too steep if the andalusite–kyanite and kyanite–sillimanite P–T lines are assumed well known. The temperature at zero pressure is $770\,°C$ instead of being greater than $800\,°C$ (Richardson et al. 1969). The enthalpy of sillimanite is only $1\cdot6$ kJ different from the calorimetric determination of Charlu et al. (1975). It is consistent with the position of the muscovite–quartz–sillimanite–sanidine–H_2O reaction determined by Chatterjee and Johannes (1974).

10 Enthalpy of formation of paragonite calculated from the paragonite–albite–corundum–H_2O and paragonite–quartz–albite–andalusite–H_2O reactions of Chatterjee (1970) and Ivanov and Gusynin (1970). The $2\cdot5$ kJ adjustment to the enthalpy of formation of andalusite is consistent with the relative positions of these two reactions.

11 Enthalpy of formation of pyrophyllite calculated from experiments on the pyrophyllite–diaspore–andalusite–H_2O and pyrophyllite–andalusite–quartz–H_2O reactions of Haas and Holdaway (1973). The $2\cdot5$ kJ adjustment to the enthalpy of formation of andalusite is confirmed by the relative positions of these two reactions.

12 Enthalpies of formation of orthoenstatite and talc calculated from experiments on the talc–enstatite–quartz–H_2O and talc–forsterite–enstatite–H_2O reactions of Chernovsky (1974). These results are consistent with the positions of the talc–magnesite–forsterite–H_2O–CO_2 and talc–quartz–magnesite–H_2O–CO_2 reactions determined by Johannes (1969). The literature calorimetric data for $MgSiO_3$ are for clinoenstatite, apart from the high temperature enthalpy of formation value given by Charlu et al. (1975). This value cannot be converted to a (1,298) value in the absence of the appropriate heat capacity data.

13 Enthalpy of formation of clinochrysotile calculated from experiments on the chrysotile–brucite–forsterite–H_2O and chrysotile–forsterite–talc–H_2O reactions of Chernovsky (1976).

14 Enthalpy of formation of anorthite calculated from experiments on the muscovite–calcite–quartz–sanidine–anorthite–CO_2–H_2O reaction by Hewitt (1973). This is disturbingly different from the value suggested by Thompson (1974), $-4247\cdot6$ kJ, which is however grossly inconsistent with the experimentally determined position of this reaction.

15 Enthalpy of formation of kaolinite calculated from experiments on the kaolinite–quartz–pyrophyllite–H_2O reaction by Thompson (1970). This value, $-4121\cdot0$ kJ, is within the uncertainty on the Robie and Waldbaum (1968) value after correction for the new gibbsite enthalpy of formation (Thompson 1974), $-4122\cdot8\pm4\cdot4$ kJ.

16 Enthalpy of formation of margarite calculated from experiments on the margarite–corundum–anorthite–H_2O reaction by Chatterjee (1974).

17 Enthalpy of formation of dolomite calculated from experiments on the dolomite–quartz–talc–calcite–H_2O–CO_2 reaction by Gordon and Greenwood (1970).
18 Enthalpy of formation of diopside calculated from experiments on the dolomite–quartz–diopside–CO_2 reaction by Slaughter *et al.* (1975). This value, -3205 kJ, is 3.5 kJ outside the uncertainty on the calorimetric value, -3198.6 ± 2.9 kJ (Navrotsky and Coons 1976). The discrepancy could arise because this value was calculated from a high temperature calorimetric value using heat capacity data of uncertain validity, or because the entropy value used for diopside in the equilibrium calculations is incorrect.
19 Enthalpy of formation of tremolite calculated from experiments on the tremolite–calcite–quartz–diopside–CO_2–H_2O,   tremolite–quartz–calcite–talc–H_2O–CO_2,   and tremolite–calcite–dolomite–quartz–H_2O–CO_2 reactions by Slaughter *et al.* (1975). This value is consistent with the position of the tremolite–diopside–enstatite–quartz–H_2O reaction of Boyd (1959)—it suggests that the thermodynamic data for the set quartz–calcite–dolomite–talc–tremolite–diopside–enstatite–forsterite are now reasonably well constrained.
20 Enthalpy of formation of grossular calculated from experiments on the grossular–quartz–anorthite–wollastonite reaction by Boettcher (1970) and Newton (1966). This value is consistent with the position of the anorthite–kyanite–grossular–quartz reaction as determined by Hays (1967) and Hariya and Kennedy (1968).
21 Enthalpy of formation of zoisite calculated from experiments on the zoisite–quartz–anorthite–grossular–H_2O reaction by Newton (1966) and Boettcher (1970). This value is consistent with the position of the zoisite–anorthite–quartz–H_2O reaction determined by Newton (1966). It is suggested that little error will arise by using these data for clinozoisite.
22 Enthalpy of formation of phlogopite calculated from experiments on the phlogopite–calcite–quartz–tremolite–sanidine–H_2O–CO_2 reaction by Hewitt (1975). This is inconsistent with the phlogopite–calcite–dolomite–sanidine–H_2O–CO_2 reaction of Puhan and Johannes (1974) by -20 K.
23 Enthalpy of formation of clinochlore calculated from experiments on the reaction chlorite–muscovite–phlogopite–kyanite–quartz–H_2O by Siefert (1970). This value was calculated assuming that the chlorite in the experiment was close in composition to clinochlore. If the same assumption is made for the chlorite–forsterite–enstatite–spinel–H_2O reaction as determined by Bird and Fawcett (1973), then this value places this reaction at about 25 K too high a temperature.

References to Notes for Table A.3
(Abbreviations: *JP = Journal of Petrology; CMP = Contributions to Mineralogy and Petrology; AM = American Mineralogist; AJS = American Journal of Science; JG = Journal of Geology; GCA = Geochimica et Cosmochimica Acta; CIWY = Carnegie Institute of Washington, Yearbook; SMPM = Schweiz. Min. Pet. Mitt.; GSAAP = Geological Society of America Abstract Programs.*)

Barnes, H. L. and Ernst, W. G., 1963, *AJS*, **261**, 129–50.
Bird, G. W. and Fawcett, J. J., 1973, *JP*, **14**, 415–28.
Boettcher, A. L., 1970, *JP*, **11**, 337–9.
Boyd, F. R., 1959. In P. H. Abelson (Ed.) *Researches in Geochemistry*, Vol. 1.
Charlu, T. V., Newton, R. C. and Kleppa, O. J., 1975, *GCA*, **39**, 1487–97.
Chatterjee, N. D., 1970, *CMP*, **27**, 244–57.
Chatterjee, N. D., 1974, *SMPM*, **54**, 753–67.
Chatterjee, N. D. and Johannes, W., 1974, *CMP*, **48**, 89–114.
Chernovsky, J. V., 1974, *GSAAP*, **6**, 687.
Chernovsky, J. V., 1976, *AM*, **61**, 1145–55.
Gordon, T. M. and Greenwood, H. J., 1970, *AJS*, **268**, 225–42.
Greenwood, H. J., 1967, *AM*, **52**, 1667–80.
Haas, H., 1972, *AM*, **57**, 1375–85.
Haas, H. and Holdaway, M. J., 1974, *AJS*, **273**, 449–61.
Hariya, Y. and Kennedy, G. C., 1968, *AJS*, **266**, 193–203.
Harker, R. I. and Tuttle, O. F., 1955, *AJS*, **253**, 209–24.
Harker, R. I. and Tuttle, O. F., 1956, *AJS*, **254**, 239–57.
Hays, J. F., 1967, *CIWY*, **65**, 234–9.

Hewitt, D. A., 1973, *AM*, **58**, 785–91.
Hewitt, D. A., 1975, *AM*, **60**, 391–7.
Hunt, J. A. and Kerrick, D. M., 1977, *GCA*, **41**, 271–88.
Ivanov, I. P. and Gusynin, V. F., 1970, *Geochemistry International*, **7**, 578–87.
Johannes, W., 1969, *AJS*, **267**, 1083–104.
Newton, R. C., 1966, *AJS*, **264**, 204–22.
Puhan, D. and Johannes, W., 1974, *CMP*, **48**, 23–31.
Richardson, S. W., Gilbert, M. C., and Bell, P. M., 1969, *AJS*, **267**, 259–72.
Siefert, F., 1970, *JP*, **11**, 73–101.
Slaughter, J., Kerrick, D. M., and Wall, V. J., 1975, *AJS*, **275**, 143–62.
Thompson, A. B., 1970, *AJS*, **268**, 454–8.
Thompson, A. B., 1974, *CMP*, **48**, 123–36.

Table A.4 F_{H_2O} and F_{CO_2} as coefficients a' and b' (where $F = a' + b'T$) tabulated for temperature ranges for a range of pressures. The volume integral part of F for H_2O is from Burnham *et al.* (1969), for CO_2 is from the Redlich–Kwong equation of state using coefficients given in Holloway (1976).

H_2O						

Temperature range/K	Pressure/kbar					
	2	4	6	8	10	
473–573	−35·7	−33·2	−30·9	−28·4	−25·9	a'/kJ
	0·1069	0·1094	0·1118	0·1134	0·1148	b'/kJ K^{-1}
573–673	−33·2	−31·4	−29·1	−26·7	−24·3	
	0·1026	0·1063	0·1086	0·1105	0·1120	
673–773	−30·7	−29·5	−27·5	−25·3	−22·9	
	0·0988	0·1033	0·1062	0·1083	0·1100	
773–873	−27·7	−27·2	−25·6	−23·5	−21·2	
	0·0949	0·1004	0·1037	0·1060	0·1077	
873–973	−25·7	−26·3	−25·0	−23·0	−20·9	
	0·0927	0·0994	0·1030	0·1055	0·1074	
973–1073	−23·4	−25·1	−24·1	−22·3	−20·2	
	0·0902	0·0982	0·1020	0·1047	0·1067	
1073–1173	−21·7	−23·4	−22·5	−20·9	−19·0	
	0·0887	0·0966	0·1006	0·1034	0·1055	
1173–1273	−19·8	−21·2	−19·5	−18·0	−16·9	
	0·0871	0·0947	0·0980	0·1009	0·1038	

CO_2						

Temperature range/K	Pressure/kbar					
	2	4	6	8	10	
473–573	−5·9	−0·8	4·8	10·5	16·3	a'/kJ
	0·0793	0·0855	0·0889	0·0914	0·0993	b'/kJ K^{-1}
573–673	−5·0	−0·1	5·5	11·2	17·0	
	0·0778	0·0842	0·0877	0·0902	0·0921	
673–773	−4·5	0·4	5·9	11·6	17·4	
	0·0771	0·0835	0·0871	0·0896	0·0915	
773–873	−4·2	0·6	6·1	11·8	17·6	
	0·0767	0·0832	0·0868	0·0893	0·0913	
873–973	−4·3	0·7	6·1	11·8	17·6	
	0·0768	0·0831	0·0868	0·0893	0·0913	
973–1073	−4·4	0·5	6·0	11·7	17·6	
	0·0769	0·0833	0·0869	0·0894	0·0913	
1073–1173	−4·7	0·4	5·9	11·6	17·4	
	0·0772	0·0834	0·0870	0·0895	0·0914	
1173–1273	−5·1	0·0	5·6	11·3	17·1	
	0·0775	0·0837	0·0873	0·0898	0·0917	

References

Robie, R. A. and Waldbaum, D. R., 1968. Thermodynamic properties of minerals and related substances at 298·15 K (25·0 °C) and one atmosphere (1·013 bars) pressure and at higher temperatures. *Bull. U.S. Geol. Surv.* **1259**.

Clark, S. P. (ed.), 1966. Handbook of physical constants. *Mem. Geol. Soc. Am.*, **97**.

Kelley, K. K., 1960. Contributions to the data on theoretical metallurgy. XIII. High-temperature heat-content, heat-capacity and entropy data for the elements and inorganic compounds. *Bull. U.S. Bur. Mines* **584**.

Burnham, C. W., Holloway, J. R. and Davis, N. F., 1969. Thermodynamic properties of water to 1000 °C and 10,000 bars. *Spec. Pap. Geol. Soc. Am.* **132**.

Charlu, T. V., Newton, R. C. and Kleppa, O. J., 1975. Enthalpies of formation at 970 K of compounds in the system MgO-Al$_2$O$_3$-SiO$_2$ from high temperature solution calorimetry. *Geochim. Cosmochim. Acta* **39**, 1487–1497.

Tardy, Y. and Garrels, R. M., 1974. A method of estimating the Gibbs energies of formation of layer silicates. *Geochim. Cosmochim. Acta* **38**, 1101–1116.

Smith, D. 1971. Stability of the assemblage Fe-rich orthopyroxene-olivine-quartz. *Am. Jour. Sci.* **271**, 370–382.

Robie, R. A., Bethke, P. M. and Beardsley, K. M., 1967. Selected X-ray crystallographic data, molar volumes, and densities of minerals and related substances. *U.S. Geol. Surv. Bull.* **1248**.

Holloway, J. R., 1977. Fugacity and activity of molecular species in supercritical fluids. *in* Fraser, D.G. (ed.) *Thermodynamics in Geology.* D. Reidel, Dordrecht.

Appendix B

A Maths Refresher

Some formulae and relationships

Re-arranging $y = \ln x$ gives $x = e^y = \exp(y)$ where $e = 2{\cdot}7183$ and $\ln$ is a logarithm to the base e (natural logarithm).

Re-arranging $y = \log x$ gives $x = 10^y$, where log is a logarithm to the base 10 (common logarithm).

Equivalence

$\ln x = 2{\cdot}3026 \log x$

Manipulation of ln

$\ln x^a = a \ln x$

$\ln xy = \ln x + \ln y$

$\ln \dfrac{x}{y} = \ln x - \ln y$

Manipulation of exp

$e^x e^y = e^{x+y}$

$e^x / e^y = e^{x-y}$

$e^{xy} = e^{xy}$

Sundries

Sum $\qquad \sum x_i = x_1 + x_2 + x_3 + x_4 + \cdots$

Product $\quad \prod x_i = x_1 . x_2 .. x_3 . x_4 \cdots$

Differentiation

This means the slope on a graph (figure B.1). The following are useful formulae, using a and b as constants, $f(x)$ and $g(x)$ both mean a function of x:

$$y = ax^b \qquad \frac{dy}{dx} = abx^{b-1}$$

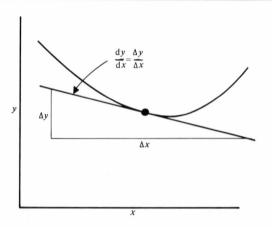

B.1 Diagram showing how differentiation can be performed graphically using the gradient of the tangent to the line at the x of interest.

For example

$$y = 3x^5 \qquad \frac{dy}{dx} = 15x^4$$

$$y = \frac{2}{x^3} = 2x^{-3} \qquad \frac{dy}{dx} = -6x^{-4} = -\frac{6}{x^4}$$

$$y = \ln x \qquad \frac{dy}{dx} = \frac{1}{x} \quad \text{and} \quad y = e^x \qquad \frac{dy}{dx} = e^x$$

$$y = f(z) \quad \text{and} \quad z = g(x) \qquad \frac{dy}{dx} = \frac{df}{dz} \cdot \frac{dz}{dx}$$

For example

$$y = \ln(x^3) \qquad \frac{dy}{dx} = \frac{3}{x}$$

$$y = \exp(x^4) \qquad \frac{dy}{dx} = 4x^3 \exp(x^4)$$

$$y = f(x)g(x) \qquad \frac{dy}{dx} = f(x) \cdot \frac{dg(x)}{dx} + g(x) \cdot \frac{df(x)}{d)x}$$

For example

$$y = x^2 \ln x \qquad \frac{dy}{dx} = x^2 \cdot \frac{1}{x} + 2x \cdot \ln x = x(1 + 2 \ln x)$$

$$y = \frac{f(x)}{g(x)} \qquad \frac{dy}{dx} = \frac{g(x) \cdot \dfrac{df(x)}{dx} - f(x) \cdot \dfrac{dg(x)}{dx}}{(g(x))^2}$$

For example

$$y = \frac{x^2}{1-x} \qquad \frac{dy}{dx} = \frac{(1-x)2x - x^2(-1)}{(1-x)^2} = \frac{1+x}{1-x}$$

Partial differentiation

If $y = f(xz)$, then

$$\left(\frac{\partial y}{\partial x} \right)_z$$

means that y is differentiated with respect to x holding z constant; so z is treated just like a and b (or 1 and 5) in the above examples (see figure B.2).

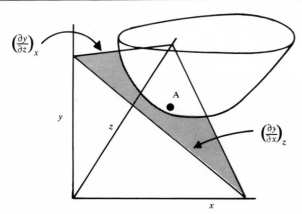

B.2   Diagram showing how partial differentiation can be performed graphically using the gradient of the tangent plane (shaded) in each direction for the x and z of interest on a bowl-shaped surface.

For example

$$y = x^2 z^3 \qquad \left(\frac{\partial y}{\partial x} \right)_z = 2xz^3 \qquad \left(\frac{\partial y}{\partial z} \right)_x = 3x^2 z^2$$

All the above methods are applicable to partial differentiation.

Integration

This is effectively the 'reverse' of differentiation because

$$y = \frac{d}{dx} \int y \, dx$$

An integral between limits x_1 and x_2 is equal to the area under the curve in a plot of y against x (figure B.3). The only necessary equation here is:

$$\int_{x_1}^{x_2} ax^b \, dx = \frac{a}{b+1} x_2^{b+1} - \frac{a}{b+1} x_1^{b+1} = \frac{a}{b+1} (x_2^{b+1} - x_1^{b+1})$$

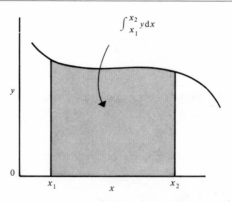

B.3 Diagram showing how integration can be performed graphically by determining the area under the curve between the x values of interest.

For example

$$\int_2^5 2x^2 \, dx = \tfrac{2}{3}(5^3 - 2^3) = 78$$

Line through two points

The equation of a line through two points (x_1, y_1) and (x_2, y_2) (see figure B.4) is

$$\frac{y - y_1}{y_2 - y_1} = \frac{x - x_1}{x_2 - x_1}$$

Re-arranging:

$$y = \underbrace{\left(\frac{y_2 - y_1}{x_2 - x_1}\right)}_{\text{slope}} x + \underbrace{\left(\frac{x_2 y_1 - x_1 y_2}{x_2 - x_1}\right)}_{\text{intercept at } x = 0}$$

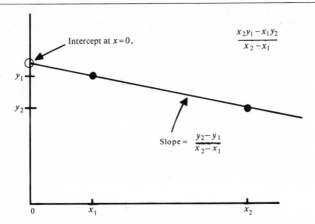

B.4   Diagram illustrating the algebraic method of putting a straight line through two points, (x_1, y_1) and (x_2, y_2).

Solving a set of simultaneous equations

The need for solving sets of simultaneous equations arises in writing balanced reactions in more complicated systems. The method (Gaussian elimination) is illustrated by the following example:

$$6a + 5b + 2c + 3d = 38$$
$$5a + 2b + 3c + 4d = 41$$
$$2a + 3b + 4c + 2d = 29$$
$$3a + 4b + c + 5d = 38$$

We start by dividing the first row through by the coefficient of the first variable in that row (i.e. 6). So the first row becomes:

$$a + \tfrac{5}{6}b + \tfrac{2}{6}c + \tfrac{3}{6}d = \tfrac{38}{6} \tag{B.1}$$

The succeeding rows are divided through by the coefficient in the first column on that row. So rows 2–4 become:

$$a + \tfrac{2}{5}b + \tfrac{3}{5}c + \tfrac{4}{5}d = \tfrac{41}{5}$$
$$a + \tfrac{3}{2}b + \tfrac{4}{2}c + \tfrac{2}{2}d = \tfrac{29}{2}$$
$$a + \tfrac{4}{3}b + \tfrac{1}{3}c + \tfrac{5}{3}d = \tfrac{38}{3}$$

Now the first row is subtracted from each of the succeeding rows. Thus these rows become:

$$-\tfrac{13}{30}b + \tfrac{8}{30}c + \tfrac{9}{30}d = \tfrac{56}{30}$$
$$\tfrac{2}{3}b + \tfrac{5}{3}c + \tfrac{1}{2}d = \tfrac{49}{6}$$
$$\tfrac{1}{2}b + \phantom{\tfrac{5}{3}c +} \tfrac{7}{6}d = \tfrac{19}{3}$$

Now, the first line of this set can be treated in the same way as the first line of the original problem. Dividing through by $-13/30$, the first row becomes:

$$b - \tfrac{8}{13}c - \tfrac{9}{13}d = -\tfrac{56}{13} \tag{B.2}$$

The succeeding rows are divided through by their first coefficients:

$$b + \tfrac{5}{2}c + \tfrac{3}{4}d = \tfrac{49}{4}$$
$$b + \phantom{\tfrac{5}{2}c +} \tfrac{7}{3}d = \tfrac{38}{3}$$

Subtracting the new first row from these:

$$\tfrac{81}{26}c + \tfrac{75}{52}d = \tfrac{861}{52}$$
$$\tfrac{8}{13}c + \tfrac{118}{39}d = \tfrac{662}{39}$$

Treating this again as a new problem and dividing through the first line by the first coefficient:

$$c + \tfrac{25}{54}d = \tfrac{287}{54} \tag{B.3}$$

and dividing through the second row by the first coefficient:

$$c + \tfrac{59}{12}d = \tfrac{331}{12}$$

Subtracting and simplifying:

$$d = 5 \tag{B.4}$$

Collecting the first lines of each successive 'problem' ((B.1)–(B.4))

$$a + \tfrac{5}{6}b + \tfrac{1}{3}c + \tfrac{1}{2}d = \tfrac{19}{3}$$
$$b - \tfrac{8}{13}c - \tfrac{9}{13}d = -\tfrac{56}{13}$$
$$c + \tfrac{25}{54}d = \tfrac{287}{54}$$
$$d = 5$$

Substituting the last line into the next to last:

$$c = 3$$

Substituting the values of c and d into the second equation:

$$b = 1$$

Substituting the values of b, c and d into the first equation gives:

$$a = 2$$

Having obtained the solution, it is always worthwhile checking that the resulting values give the appropriate right-hand side value for each of the original equations.

Appendix C
Standard States for Fluids

Complications arise in devising standard states for end-members of fluids. The reason for this is that the Gibbs energy of a gas is a simple function of P–T at very low pressures and so the properties of the gas are measured under these conditions. At very low pressures, as the volume becomes large, the atoms or molecules interact little with each other, and the following relation, the gas law, holds:

$$PV = RT \tag{C.1}$$

A fluid which obeys the gas law is called a perfect gas (or fluid). The change of Gibbs energy with pressure ((A.1) on p. 229) for a perfect gas is:

$$\int_{P_1}^{P_2} V \, dP = \int_{P_1}^{P_2} \frac{RT}{P} \, dP = RT \ln \frac{P_2}{P_1} \tag{C.2}$$

and so, for a perfect gas, applying this to (3.9) gives:

$$\mu_{1A} = G_{1A}(1, T) + RT \ln x_{1A} \gamma_{1A} P \tag{C.3}$$

However, no fluid is perfect over a geologically useful range of P–T, so unfortunately this simple equation is not sufficient.

However, even real gases become perfect at sufficiently low pressures. Consider that we know the Gibbs energy of the gas at some low pressure,

P' at which the gas is perfect. Then, analogously to (3.9):

$$\mu_{1A} = G_{1A}(P', T) + \int_{P'}^{P} V_{1A}\, dP + RT \ln x_{1A}\gamma_{1A} \tag{C.4}$$

The Gibbs energy of the hypothetical perfect gas at 1 bar, $G'(1, T)$, is:

$$G'_{1A}(1, T) = G_{1A}(P', T) + \int_{P'}^{1} \frac{RT}{P}\, dP = G_{1A}(P', T) + RT \ln \frac{1}{P'}$$

Re-arranging and combining with (C.4)

$$\mu_{1A} = G'_{1A}(1, T) + RT \ln P' + \int_{P'}^{P} V_{1A}\, dP + RT \ln x_{1A}\gamma_{1A} \tag{C.5}$$

$$= G'_{1A}(1, T) + RT \ln \left[x_{1A}\gamma_{1A}P' \exp \left(\int_{P'}^{P} \frac{V_{1A}}{RT}\, dP \right) \right] \tag{C.6}$$

In (C.5), $G'(1, T)$ is the Gibbs energy of the hypothetical perfect gas at 1 bar and T, $RT \ln P'$ takes this value down pressure to a pressure, P', at which the actual gas is perfect, and the volume integral then takes the result up to the pressure of interest, P. The standard state to use then is standard state 1, pure (actual) end-member at the $P–T$ of interest. The tabulation of data is however connected with the above, in that the tabulated 1 bar data (for example in appendix A) is for the hypothetical perfect gas, although the term which takes this up pressure includes the correction for the hypothetical nature of the 1 bar data. This term is included in the F term (see appendix A, p. 246).

In (C.6) the exponential term is usually written in terms of the fugacity, f, which can be defined as:

$$\frac{f_{1A}(P, T)}{f_{1A}(P', T)} = \exp \left(\int_{P'}^{P} \frac{V_{1A}}{RT}\, dP \right)$$

with the proviso that the fugacity is equal to the pressure when the gas is perfect (i.e. at P'), so $f_{1A}(P', T) = P'$, thus:

$$\frac{f_{1A}(P, T)}{P'} = \exp \left(\int_{P'}^{P} \frac{V_{1A}}{RT}\, dP \right) \tag{C.7}$$

Substituting in (C.6), the P' terms cancel, and:

$$\mu_{1A} = G'_{1A}(1, T) + RT \ln x_{1A}\gamma_{1A}f_{1A} \tag{C.8}$$

The fugacity, by comparison with (C.3), can be viewed as a thermodynamic pressure for gases. An alternative representation involves the fugacity coefficient, Γ, which is defined as $\Gamma = f/P$, which has the useful property of being equal to one when the gas is perfect. As gases tend to approach perfect behaviour towards high temperature at a particular pressure, and towards low pressure at a particular temperature, then Γ is particularly useful for tabulating gas data because Γ varies much less rapidly with temperature and pressure and also has a smaller magnitude than the fugacity. Substituting $f = P\Gamma$ into (C.8) gives:

$$\mu_{1A} = G'_{1A}(1, T) + RT \ln x_{1A} \gamma_{1A} P \Gamma_{1A} \tag{C.9}$$

which can also be written in terms of the partial pressure, $p_{1A} = Px_{1A}$, as:

$$\mu_{1A} = G'_{1A}(1, T) + RT \ln p_{1A} \gamma_{1A} \Gamma_{1A} \tag{C.10}$$

In these three equivalent expressions it is important to realize that f_{1A} and Γ_{1A} both refer to the gas made up solely of end-member 1 at the temperature and pressure of interest, whereas x_{1A}, γ_{1A} and p_{1A} all refer to the composition of the actual gas at the temperature and pressure of interest.

Equation (C.8) can be related to a new standard state (standard state 4(a)) of the pure end-member as a hypothetical perfect gas at 1 bar and the temperature of interest. So:

$$\mu^{\circ}_{1A} = G'_{1A}(1, T) \quad \text{and} \quad a_{1A} = x_{1A} \gamma_{1A} f_{1A} = p_{1A} \gamma_{1A} \Gamma_{1A}$$

However, it is worthwhile investigating the actual difference between the properties of real gases and hypothetical perfect gases at 1 bar. At $0 \cdot 01$ bar H_2O is a perfect gas. Burnham *et al.* (1969) give the Gibbs energy of real H_2O gas at $0 \cdot 01$ bar and 1 bar. From (C.2) the difference between these should be $RT \ln 0 \cdot 01$ if H_2O is a perfect gas at 1 bar. The discrepancy is trivial, being less than $0 \cdot 05$ kJ above $300\,^{\circ}C$. Thus the above convolutions are unnecessary as the Gibbs energy of the real gas at 1 bar is effectively the same as the Gibbs energy of the perfect gas at 1 bar. Thus (C.5) can be rewritten as:

$$\mu_{1A} = G_{1A}(1, T) + \int_1^P V_{1A} \, dP + RT \ln x_{1A} \gamma_{1A}$$

which is standard state 1; and (C.6)–(C.8):

$$\mu_{1A} = G_{1A}(1, T) + RT \ln \left[x_{1A} \gamma_{1A} \exp \left(\int_1^P \frac{V_{1A}}{RT} \, dP \right) \right]$$

$$= G_{1A}(1, T) + RT \ln x_{1A} \gamma_{1A} f_{1A}$$

which is standard state 4, with the fugacity acting as a shorthand form for the exponential term. Thus it is not necessary in practice to stipulate that the standard state used for end-members of fluids is a hypothetical one, and this is not done in the text or in appendix A.

Solutions to problems 2

2(a) Weight proportions: Fe_3O_4, 0·794 and SiO_2, 0·206; volume proportions Fe_3O_4, 0·662 and SiO_2, 0·338.

2(b) See figure D.1.

2(c) As PY contains seven molecules of SiO_2, MgO and Al_2O_3, while KY
contains only two; the proportion of the garnet in the rock will be
less than that of kyanite. As in WE2(c), the ratio of proportions of
PY and KY is PY/KY = 3·5. PY + KY must equal 1. Solving gives
the mole proportion of PY = 0·222, and KY = 0·778.

2(d) The system involves three oxides so four phases are required to
write a balanced reaction. As there are five phases, five reactions
can be written, with one phase omitted in each.

$$[PY] \quad 2Mg_2SiO_4 + Al_2SiO_5 = MgAl_2O_4 + 3MgSiO_3$$
$$[KY] \quad Mg_2SiO_4 + Mg_3Al_2Si_3O_{12} = 4MgSiO_3 + MgAl_2O_4$$
$$[SP] \quad Mg_2SiO_4 + MgSiO_3 + Al_2SiO_5 = Mg_3Al_2Si_3O_{12}$$
$$[FO] \quad 5MgSiO_3 + MgAl_2O_4 + Al_2SiO_5 = 2Mg_3Al_2Si_3O_{12}$$
$$[EN] \quad 3Mg_3Al_2Si_3O_{12} + MgAl_2O_4 = 5Mg_2SiO_4 + 4Al_2SiO_5$$

2(e) See figure D.2(a).

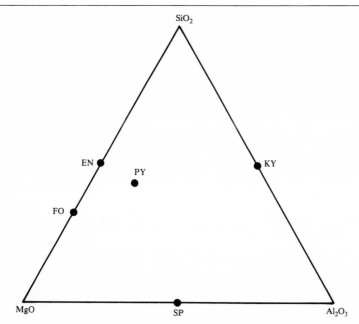

D.1 Solution to problem 2(b).

2(f) See figure D.2(b).

2(g) For (a) there are five reactions:

[GR] $SiO_2 + Al_2SiO_5 + 2Ca_2Al_3Si_3O_{12}(OH)$
$$= 4CaAl_2Si_2O_8 + H_2O$$

[Z] $SiO_2 + 2Al_2SiO_5 + Ca_3Al_2Si_3O_{12} = 3CaAl_2Si_2O_8$

[S] $SiO_2 + 4Ca_2Al_3Si_3O_{12}(OH) = 5CaAl_2Si_2O_8 + Ca_3Al_2Si_3O_{12}$
$$+ 2H_2O$$

[AN] $SiO_2 + 5Al_2SiO_5 + 4Ca_3Al_2Si_3O_{12} + 3H_2O$
$$= 6Ca_2Al_3Si_3O_{12}(OH)$$

[Q] $Ca_3Al_2Si_3O_{12} + CaAl_2Si_2O_8 + Al_2SiO_5 + H_2O$
$$= 2Ca_2Al_3Si_3O_{12}(OH)$$

For (b), [Q] is not relevant because Q is considered to be present in excess.

2(h) $114CaAl_2Si_2O_8 + 35KMg_3AlSi_3O_{10}(OH)_2 + 3Ca_2Mg_5Si_8O_{22}(OH)_2$
$$+ 88H_2O$$

$$= 35KAlSi_3O_8 + 60Ca_2Al_3Si_3O_{12}(OH) + 24Mg_5Al_2Si_3O_{10}(OH)_8$$

Solutions to problems 3

3(a)   (a) When $x_{1A} \to 1$; $x_{2A} \to 0$ and $\gamma_{1A} \to 1$ and $\gamma_{2A} \to \exp(w_2/RT)$;
also $x_{2A} \to 1$; $x_{1A} \to 0$ and $\gamma_{2A} \to 1$ and $\gamma_{1A} \to \exp(w_1/RT)$.

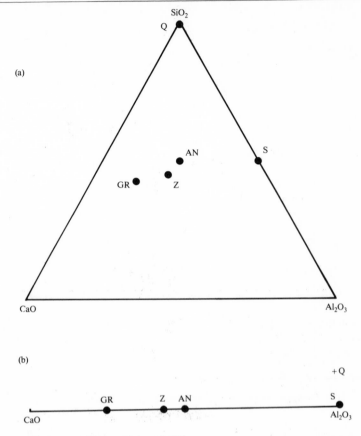

(a)

(b)

D.2 Solution to problems 2(e) and 2(f).

(b) If $w_1 = w_2 = w$ then

$$RT \ln \gamma_{1A} = wx_{2A}^2$$

$$RT \ln \gamma_{2A} = wx_{1A}^2$$

which are the regular model equations.

3(b) G_A simplifies to:

$$G_A = RT(x_{1A} \ln x_{1A} + x_{2A} \ln x_{2A}) + x_{1A}x_{2A}(x_{2A}w_1 + x_{1A}w_2)$$

(using $G_{1A} = G_{2A} = 0$). Substituting into this in the manner of WE 3(d) gives figure D.3.

3(c) Using Q for the phase quartz and s for SiO_2; S for sillimanite and a for Al_2SiO_5; PL for plagioclase and an for $CaAl_2Si_2O_8$; Z for zoisite and z for $Ca_2Al_3Si_3O_{12}(OH)$; G for garnet and gr for $Ca_3Al_2Si_3O_{12}$; and F for fluid and w for H_2O. Each equilibrium relation can be written:

$$\Delta G^\circ + RT \ln K = 0$$

Using standard state 1 for each end-member:

for [GR]

$$\Delta G^\circ = 4G_{\text{an,PL}} + G_{\text{w,F}} - 2G_{\text{z,Z}} - G_{\text{a,S}} - G_{\text{s,Q}}$$

$$K = \frac{a_{\text{an,PL}}^4 a_{\text{w,F}}}{a_{\text{s,Q}} a_{\text{a,S}} a_{\text{z,Z}}^2}$$

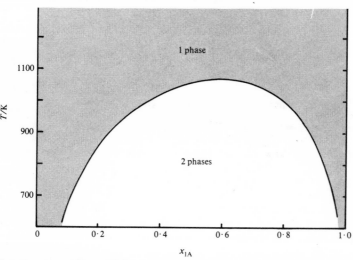

D.3   Solution to problem 3(b).

for [Z]

$$\Delta G^\circ = 3G_{\text{an,PL}} - G_{\text{s,Q}} - 2G_{\text{a,S}} - G_{\text{gr,G}}$$

$$K = \frac{a_{\text{an,PL}}^3}{a_{\text{s,Q}} a_{\text{a,S}}^2 a_{\text{gr,G}}}$$

for [S]

$$\Delta G^\circ = 5G_{\text{an,PL}} + G_{\text{gr,G}} + 2G_{\text{w,F}} - G_{\text{s,Q}} - 4G_{\text{z,Z}}$$

$$K = \frac{a_{\text{an,PL}}^5 a_{\text{gr,G}} a_{\text{w,F}}^2}{a_{\text{s,Q}} a_{\text{z,Z}}^4}$$

for [AN]

$$\Delta G^\circ = 6G_{\text{z,Z}} - G_{\text{s,Q}} - 5G_{\text{a,S}} - 4G_{\text{gr,G}} - 3G_{\text{w,F}}$$

$$K = \frac{a_{\text{z,Z}}^6}{a_{\text{s,Q}} a_{\text{a,S}}^5 a_{\text{z,Z}}^4 a_{\text{w,F}}^3}$$

for [Q]

$$\Delta G° = 2G_{z,Z} - G_{gr,G} - G_{an,PL} - G_{a,S} - G_{w,F}$$

$$K = \frac{a_{z,Z}^2}{a_{gr,G}\, a_{an,PL}\, a_{a,S}\, a_{w,F}}$$

Solutions to problems 4

4(a) The site occupancies are:

$$x_{Si,T} = 0.9819 \qquad x_{Al,T} = 0.0181$$

$$x_{Al,M1} = 0.0018 \qquad x_{Cr,M1} = 0.0123 \qquad x_{Fe3,M1} = 0.0106$$

$$x_{Ca,M2} = 0.0084$$

$$x_{Mg,M1} = 0.9208 \qquad x_{Fe,M1} = 0.0545$$

$$x_{Mg,M2} = 0.9362 \qquad x_{Fe,M2} = 0.0554$$

The mole fractions, by mixing on sites:

$$x_{Fe_2Si_2O_6} = 0.00291$$

$$x_{Mg_2Si_2O_6} = 0.831$$

$$x_{CaMgSi_2O_6} = 0.00746$$

$$x_{CaCrAlSiO_6} = 7.35 \times 10^{-7}$$

$$x_{Mg_{1.5}AlSi_{1.5}O_6} = 0.0308$$

4(b) The site occupancies are:

$$x_{Si,Ti} = 0.95 \qquad x_{Al,T1} = 0.05$$

$$x_{OH,V} = 0.85 \qquad x_{F,V} = 0.15$$

$$x_{o,A} = 0.9 \qquad x_{Na,A} = 0.1$$

$$x_{Na,M4} = 0.9 \qquad x_{Ca,M4} = 0.1$$

$$x_{Al,M2} = 0.95$$

$$x_{Mg,M13} = 0.839 \qquad x_{Fe,M13} = 0.161$$

$$x_{Mg,M2} = 0.0419 \qquad x_{Fe,M2} = 0.00806$$

The mole fractions, by mixing on sites:

$$x_{Ca_2Mg_5Si_8O_{22}(OH)_2} = 5.49 \times 10^{-6}$$

$$x_{Na_2CaMg_5Si_8O_{22}(OH)_2} = 2.20 \times 10^{-5}$$

$$x_{Na_2Mg_3Al_2Si_8O_{22}(OH)_2} = 0.229$$

$x_{Mg_7Si_8O_{22}(OH)_2}$ is undefined because $x_{Mg,M4} = 0$

$$x_{NaCa_2Mg_4AlSi_6Al_2O_{22}(OH)_2} = 2.45 \times 10^{-6}$$

$$x_{Na_2Mg_3Al_2Si_8O_{22}(F)_2} = 0.00712$$

4(c) The mole fractions are:

(1) $x_{Mg_2SiO_4,OL} = 0 \cdot 16$

$x_{Fe_2SiO_4,OL} = 0 \cdot 36$

(2) $x_{Mn_3Al_2Si_3O_{12},G} = 6 \cdot 00 \times 10^{-5}$

$x_{Fe_3Al_2Si_3O_{12},G} = 2 \cdot 03 \times 10^{-4}$

$x_{Ca_3Fe_2Si_3O_{12},G} = 0 \cdot 136$

(3) $x_{NaAlSi_3O_8,PL} = 0 \cdot 7$

$x_{CaAl_2Si_2O_8,PL} = 0 \cdot 1$

$x_{KAlSi_3O_8,PL} = 0 \cdot 2$

(4) $x_{Fe_5Al_2Si_3O_{10}(OH)_8,CHL} = 0 \cdot 0366$

$x_{Fe_4Al_4Si_2O_{10}(OH)_8,CHL} = 0 \cdot 0388$

Solutions to problems 5

5(a) The equilibrium relation can be re-arranged to give:

for (a) $T = \dfrac{20820}{27 \cdot 4 - \ln \dfrac{x^3}{1-x}}$ T in K $x = x_{CO_2}$

for (b) $T = \dfrac{20820 + 2500(3x^2 + x - 1)}{27 \cdot 4 + 1 \cdot 8(3x^2 + x - 1) - \ln \dfrac{x^3}{1-x}}$

The $T-x$ curves which result from substituting into these equations are given in figure D.4. Note the substantial flattening effect of the activity coefficients on the shape of the $T-x$ curve.

5(b) The equilibrium relation can be re-arranged to give:

$$\ln a_{NaCl} = -\frac{42370}{T} + 25 \cdot 5   \quad T \text{ in K}$$

Figure D.5 shows the $\ln a_{NaCl}-1/T$ diagram which results from substituting into this equation. At $\ln a_{NaCl} = -24$, nepheline is replaced by sodalite at 600 °C during cooling.

5(c) Using the same approach as WE 5(c):

$m_{KSO_4^-} = 9 \cdot 4 \times 10^{-7}$ $\dot{m}_{SO_4^{2-}} = 2 \cdot 99 \times 10^{-4}$

$m_{NaSO_4^-} = 6 \cdot 0 \times 10^{-7}$ $m_{K^+} = 3 \cdot 99 \times 10^{-4}$

$m_{Na^+} = 1 \cdot 99 \times 10^{-4}$

Note the small degree of complexing.

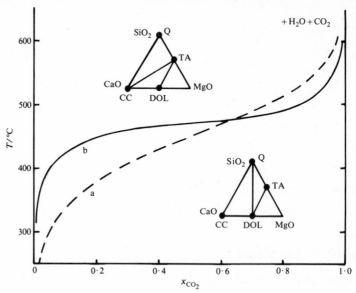

D.4 Solution to problem 5(a).

5(d) The reactions of interest with their equilibrium expressions are:

$2Cu + H_2O = Cu_2O + 2H^+ + 2e$ $0 = 18 \cdot 3 + \ln a_e + \ln a_{H^+}$

$Cu_2O + H_2O = CuO + 2H^+ + 2e$ $0 = 26 \cdot 1 + \ln a_e + \ln a_{H^+}$

$Cu = Cu^{2+} + 2e$ $0 = 13 \cdot 1 + \ln a_e \qquad + \frac{1}{2} \ln a_{Cu^{2+}}$

$2Cu^{2+} + H_2O + 2e = Cu_2O + 2H^+$ $0 = 7 \cdot 9 + \ln a_e - \ln a_{H^+} + \ln a_{Cu^{2+}}$

$Cu^{2+} + H_2O = CuO + 2H^+$ $0 = 9 \cdot 1 \qquad + \ln a_{H^+} - \frac{1}{2} \ln a_{Cu^{2+}}$

These are plotted on figure D.6.

Solutions to problems 6

6(a) The geothermometer expression is:

$$T = \frac{8200}{4 \cdot 6 - \ln \dfrac{x(1-y)}{(1-x)y}}$$

Then:

$$\left(\frac{\partial T}{\partial a}\right)_{b,x,y} = \frac{T}{a} \qquad\qquad \left(\frac{\partial T}{\partial b}\right)_{a,x,y} = -\frac{T^2}{a}$$

$$\left(\frac{\partial T}{\partial x}\right)_{a,b,y} = \frac{T^2}{ax(1-x)} \qquad\qquad \left(\frac{\partial T}{\partial y}\right)_{a,b,x} = -\frac{T^2}{ay(1-y)}$$

Using just the first two: $\sigma_T = 21$;
Using all four: $\sigma_T = 28$.

Note that if x or y approach 0 or 1, the uncertainty on T increases rapidly because of the form of the dependence of σ_T on x and y.

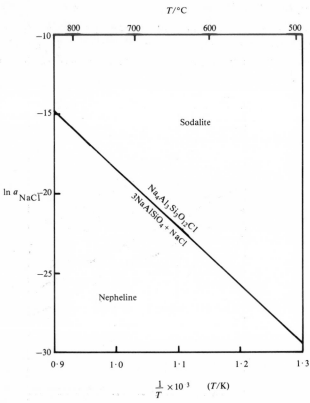

D.5   Solution to problem 5(b).

6(b)   Using the same symbols as in WE4(a), (b):

$$2x_{Na,M4} = 7 - C_{Fe} - C_{Mg} - C_{Ca} - C_{Fe3} - C_{Ti} - [C_{Al} - (8 - C_{Si})]$$

or:

$$x_{Na,M4} = \tfrac{1}{2}(15 - FH)$$

where:

$$H = \frac{W_{FeO}}{MW_{FeO}} + \frac{W_{MgO}}{MW_{MgO}} + \frac{W_{CaO}}{MW_{CaO}} + \frac{2W_{Fe_2O_3}}{MW_{Fe_2O_3}} + \frac{2W_{Al_2O_3}}{MW_{Al_2O_3}} + \frac{W_{TiO_2}}{MW_{TiO_2}}$$
$$+ \frac{W_{SiO_2}}{MW_{SiO_2}}$$

Now:

$$\frac{dx_{Na,M4}}{dW_{SiO_2}} = \frac{F}{MW_{SiO_2}}\left(\frac{HF}{N} - \frac{1}{2}\right)$$

Substituting: $F = 8.663$, $H = 1.734$. $N = 23$, $MW_{SiO_2} = 60.09$. Therefore $\sigma_{x_{Na,M4}} = 0.011$.

The value of $x_{Na,M4}$ is -0.009 ± 0.011. Therefore the presence of M4 site Na cannot be ruled out, particularly as the effect of the uncertainties on the other wt% values will increase the uncertainty on $x_{Na,M4}$.

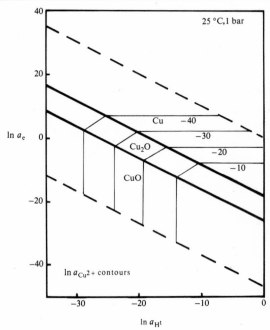

D.6 Solution to problem 5(d).

6(c) The $\Delta G°$ for the two reactions from the appendix data are:

$$\Delta G_1° = 73.5 - 0.1584T - 2.28P$$
$$\Delta G_2° = 163.2 - 0.3602T - 3.67P$$

(T in K, P in kbar)

The resulting equilibrium relations at 2 kbar are:

$$0 = 64.4 - 0.0813T + RT \ln x + (20.8 - 0.015T)(1-x)^2$$

$$0 = 173.1 - 0.2277T + RT \ln \frac{0.75x^3}{1-x} + (20.8 - 0.015T)$$
$$\times [3(1-x)^2 - x^2]$$

Substituting into these equations gives the main curves on figure D.7. Adjusting the ΔG° values by the suggested uncertainties and repeating the calculation gives the band for each reaction. The temperature of the intersection lies between 490 °C and 530 °C and the x_{CO_2} between 0·75 and 0·98.

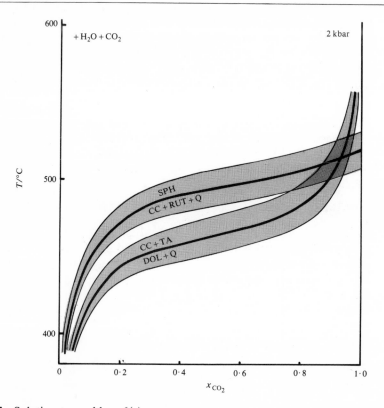

D.7   Solution to problem 6(c).

Solutions to problems 8

8(a) (1) Internal buffering: at 390 °C, FO will start to form on FO = CH + BR and the fluid will become more H_2O-rich. After an infinitesimal increase in temperature (because of the closeness to $x_{CO_2} = 0$), either CH or BR will be consumed depending on their original proportions (they are used up in equal proportions by the reaction). If BR is used up, the assemblage will develop TA at 470 °C and the fluid will become more H_2O-rich. After an infinitesimal rise in temperature, the CH will be consumed, leaving the resulting assemblage to pass to higher temperatures. If CH is used up on $FO = CH + BR$, then the assemblage will not 'see' the TA-producing reaction.

External buffering (at $x_{CO_2} = 0.0002$): the sequence of assemblages is indistinguishable from the above, although the fluid does not change composition in crossing any of the reaction lines.

(2) Internal buffering: the assemblage will move up $BR = MAG$ to the intersection at 390 °C, where FO starts to form. The reaction here is:

$$CH + 0.0009MAG + 0.9991BR = 2FO + fluid$$

BR is reacted away 1100 times quicker than MAG. Only if the proportion of MAG is less than 1/1100 of the proportion of BR will the assemblage exit along $CH + BR = FO$ and a similar sequence to (1) above will result. Otherwise, the assemblage will exit along $CH + MAG = FO$. A trivial additional amount of FO will be developed along this reaction to the intersection at 470 °C. At the intersection, the reaction is:

$$0.093MAG + 5.093CH = 6.186FO + TA + 9.278fluid$$

CH is used up 55 times faster than MAG. Therefore, unless the proportion of MAG is less than 1/55 of the proportion of CH, the assemblage will move along $TA + MAG = FO$, with the fluid becoming progressively more CO_2-rich. At 570 °C, EN appears in the assemblage at the intersection at $x_{CO_2} = 0.95$. Note that $EN + MAG$ bearing assemblages can be formed by the process of internal buffering during the metamorphism of serpentinite, but not $EN + MAG$ assemblages in the absence of TA.

External buffering (at $x_{CO_2} = 0.0003$): the same sequence as in (1) above.

Appendix E

Symbols, Units and Constants

Symbols

The main symbols and their meanings are listed below. Some symbols inevitably are used for several different purposes at different places in the text. They are defined where they occur.

a_{1A} activity of end-member 1 in phase A

C_p heat capacity at constant pressure

F function defined for H_2O and CO_2 in (A.21), (A.22); $F = a' + b'T$

f_{1A} fugacity of pure end-member 1 in the structure of phase A

G Gibbs energy

G_{1A} Gibbs energy of pure end-member 1 in the structure of phase A

H enthalpy

$h_{1A(2)}$ Henry's law constant for end-member 1 in infinite dilution in end-member 2 in phase A

I ionic strength

K equilibrium constant

m_i molality of i

n_{1A} number of moles of end-member 1 in phase A

P pressure

p_{1A} pressure of end-member 1 in phase A (partial pressure)

R gas constant, $R = 0 \cdot 0083144 \text{ kJ K}^{-1}$

S entropy

T temperature

t time

V volume

w regular model mixing parameter

x_{1A} mole fraction of end-member 1 in phase A

α coefficient of thermal expansion

γ_{1A} activity coefficient of end-member 1 in phase A

κ isothermal compressibility

μ_{1A} chemical potential of end-member 1 in phase A

(1,298) postfix for 1 bar and 298 K

(1,T)   postfix for 1 bar and T K

(P,T)   postfix for P kbar and T K

Δ prefix for the change in some property for a balanced reaction

Δ_f prefix for the change in some property for a balanced reaction forming that phase from its constituent elements

$\circ$ superscript for a standard state property; for example

        $\Delta_f G(1,298)$ = Gibbs energy of formation (from the elements) at 1 bar and 298 K

        ΔG° = standard Gibbs energy of reaction

Units

SI units are used, except for pressure, as pascals (Pa) (1 bar = 10^5 Pa) are an unwieldly unit for geological pressures.

Pressure, P: kbar (for convenience, 1 (i.e. 1 bar) is used in the postfix (1,298), rather than 0·001)

Temperature, T: degrees Kelvin, K

                  degrees Centigrade, °C

                  conversion, K = °C + 273

Volume, V: kJ kbar^{-1}

Entropy, S: kJ K^{-1}

Enthalpy, H: kJ

Gibbs energy, G: kJ

Constants

$R = 0·0083144$ kJ K^{-1}

molecular weights:

SiO_2	60·09	CaO	56·08	K_2O	94·20
TiO_2	79·90	FeO	71·85	Na_2O	61·98
Al_2O_3	101·94	MgO	40·32	H_2O	18·02
Cr_2O_3	152·02	MnO	70·94	CO_2	44·01
Fe_2O_3	159·70	NiO	74·71		

Index